M&A Deal Mechanics

From LOI to Close

Joseph Trek

Published by NobleTrex Press

For permissions and other inquiries, write to:

P.O. Box 3132, Framingham, MA 01701, USA

Contents

Introduction

It was 4:58 PM on a Friday, and the conference room smelled like stale coffee and anxiety.

We had been working on the transaction for six months. We had traded three thousand emails, reviewed four thousand documents, and argued over everything from the environmental status of a warehouse in Ohio to the definition of a "business day." But right now, none of that mattered. The only thing that mattered was a sixteen-digit federal reference number.

We were waiting for the wire.

The buyer's bank had sworn the funds were released an hour ago. The seller-a founder who had spent thirty years building this manufacturing company-was pacing the floor, checking his phone every twelve seconds. If the money didn't hit the account before the Fed wire window closed at 5:00 PM, the deal wouldn't technically close until Monday. Over a weekend, anything can happen. Remorse kicks in. A factory catches fire. A pandemic starts. Dealmakers live in terror of the "gap period," and right now, we were living in the gap.

At 5:01 PM, the phone buzzed. The number appeared. The tension in the room snapped like a rubber band. Handshakes were exchanged, stiff and exhausted. The founder looked at me, eyes red-rimmed, and asked the question that haunts every M&A professional:

"So... is that it? I just sold my life's work for a PDF confirmation?"

That moment-the collision of massive financial consequence with anticlimactic administrative procedure-is the essence of Mergers and Acquisitions. From the outside, M&A looks like a strategy game played by titans of industry in glass towers. We read headlines: *Com-*

pany X acquires Company Y for $2 Billion. It sounds clean. It sounds like a handshake.

But if you are reading this book, you suspect the truth: M&A is not a handshake. It is a machine.

It is a grinding, complex, often counterintuitive mechanical process designed to transfer risk and value from one pile of legal documents to another. It is a discipline where accounting, law, psychology, and raw negotiation leverage crash into each other. And the distance between "We have a deal" (the handshake) and "Here is your money" (the close) is filled with a thousand traps that can destroy value or kill the transaction entirely.

This book is an operating manual for that machine.

The Illusion of the Headline Price

Why do you need to understand deal mechanics? Because the headline price is almost never what the seller takes home.

Let's say you agree to sell your business for $100 million. You might mentally spend that money on a beach house or a diversified portfolio. But in the world of M&A, $100 million is just an "Enterprise Value." It is a theoretical number based on the assumption that the business is delivered with zero debt and a normal amount of cash.

But businesses are messy. They have bank loans, unpaid bonuses, and customers who haven't paid their bills yet. By the time we run that $100 million through the machinery of the "waterfall"-subtracting indebtedness, adjusting for working capital targets, holding back money for escrow, and paying the army of advisors-the "Equity Value" (the check you actually receive) might look more like $60 million.

If you don't understand the mechanics of the *bridge* between Enterprise Value and Equity Value, you aren't negotiating the deal; you are just participating in a hallucination.

This isn't just for investment bankers or private equity sharks. If you are a business owner, an aspiring corporate development officer, or a lawyer trying to bridge the gap between theory and practice, you need to speak the language of the deal. You need to understand why a "Cash-Free, Debt-Free" transaction is the standard, yet arguably the most confusing phrase in finance. You need to know why "Working Capital" is the place where buyers go to bury price cuts, and why "Indemnification" is the scariest word in a purchase agreement.

The Architecture of a Transaction

We are going to dismantle the deal process piece by piece, chronologically. We won't be using academic theories about synergy or corporate strategy. There are plenty of books on *why* companies merge. This book is about *how* they merge.

The process usually begins with a courtship. In our first chapters, we explore the landscape and the "Letter of Intent" (LOI). The LOI is a strange document. It is the engagement ring of the M&A world-it signals intent and takes the seller off the market, but it isn't a marriage license.

Many novices make the mistake of celebrating when the LOI is signed. They pop champagne because they see a big number on a piece of paper. The pros know better. The pros know that the LOI is where you have the most leverage you will ever possess. Once you sign it and grant the buyer exclusivity (the "No-Shop" clause), the power dynamic flips instantly. You are now trapped in a room with a buyer who is incentivized to find problems. We will show you how to craft an LOI that protects your leverage, distinguishing between the terms that are legally binding and the ones that are merely "agreements to agree."

The weaponization of accounting

Once the LOI is signed, the doors open to Due Diligence. If you think diligence is just an audit to check if the bank statements match the tax returns, you are in for a rude awakening.

Modern diligence is a strategic weapon. Buyers use "Quality of Earnings" (QoE) reports to reconstruct your profitability. They will argue that your EBITDA (Earnings Before Interest, Taxes, Depreciation, and Amortization) isn't real. They will claim that your one-time add-backs are actually recurring expenses. They will argue that you haven't been paying your employees enough, and therefore they need to lower the purchase price to account for the "true" cost of labor.

We will walk through the "diligence engine" not to teach you how to be an accountant, but to teach you how to defend against one. We will explore how a "finding" in diligence morphs into a price reduction, or how it gets quarantined in a "Disclosure Schedule"-a defensive document that effectively says, "I told you about this mess, so you can't sue me for it later."

The Contractual Shield

As diligence wraps up, we move to the "Definitive Agreements." This is where the lawyers earn their fees, and where business people often tune out. That is a fatal mistake.

The Purchase Agreement is not just legal boilerplate. It is a risk-allocation engine. It decides who pays the bill when things go wrong after closing.

Imagine you sell a software company. Six months later, a third party sues the buyer, claiming your code infringed on their patent. Who pays for that lawsuit? You, or the new owner? The answer lies in the dense thicket of "Representations and Warranties" and "Indemnification Caps."

We will decode this language. We will explain the modern revolution of "Rep and Warranty Insurance" (RWI), a product that has fundamentally changed dealmaking by allowing sellers to walk away with more cash at closing, shifting the risk to an insurance carrier instead of an escrow account. We will explain why the definition of

"Fraud" is often the most hotly contested paragraph in a hundred-page document.

The Space Between Signing and Closing

Finally, we will navigate the "limbo" period. In the old days, signing and closing happened in the same room on the same day. Today, regulatory hurdles (like antitrust reviews) often force a split: you sign the deal in April, but you don't close until July.

What happens in those three months? You still own the company, but you can't really run it. You are bound by "interim covenants" that prevent you from giving raises, firing executives, or changing prices without the buyer's permission. It is a governance nightmare. We will discuss how to manage this tension without breaching the contract or destroying the business morale before the keys are handed over.

A Note on the "Smart Friend" Approach

I have written this book to be read, not referenced. Finance literature has a bad habit of being dry, hiding behind acronyms to appear sophisticated. But complexity is not sophistication; clarity is.

Throughout these pages, I will speak to you like a smart friend explaining the ropes over a drink. I will use analogies. We will compare LOIs to dating, indemnification baskets to car insurance deductibles, and closing conditions to launch codes.

We will also focus heavily on the "why." Why do Private Equity firms insist on "rollover equity"? It's not just to be nice; it's to trap the founder in the boat so they don't drill a hole in the hull after they leave. Why do bankers obsess over "Net Working Capital"? Because it's the easiest place for a seller to artificially inflate cash flow right before a sale.

The Stakes

Ultimately, M&A is the point where the rubber of capitalism meets the road. It is how efficiencies are realized, how founders are rewarded for decades of risk, and how industries evolve. But it is also a zero-sum game in many respects. Every dollar of risk shifted to the buyer is a dollar of safety for the seller. Every adjustment to the purchase price is a direct transfer of wealth.

When you are in the thick of a deal, the fatigue is overwhelming. You will be answering emails at 2:00 AM. You will be fielding calls from panicked employees. You will be reading markups of contracts that seem to say the same thing in slightly different Latin phrases.

In those moments, you need a map. You need to know where you are in the lifecycle, what leverage you hold, and what is coming next.

This book is that map.

So, let's pop the hood. Let's ignore the headlines and the press releases and look at the gears grinding underneath. Let's find out what actually happens between the first handshake and the final wire transfer.

Chapter 1

The M&A Landscape and Deal Foundations

Buying or selling a company often looks like a straightforward handshake from the outside, but underneath the surface, it is a complex machine of moving parts. This chapter pulls back the curtain on that machine. We will move beyond the jargon to understand the 'who, what, and how' of a transaction-from the timeline that dictates the pace to the structural decisions that determine whether you are buying a business or just its inventory. By the end, you will understand why the price on the headline is rarely the amount that hits the bank account.

1.1 The Lifecycle of a Deal

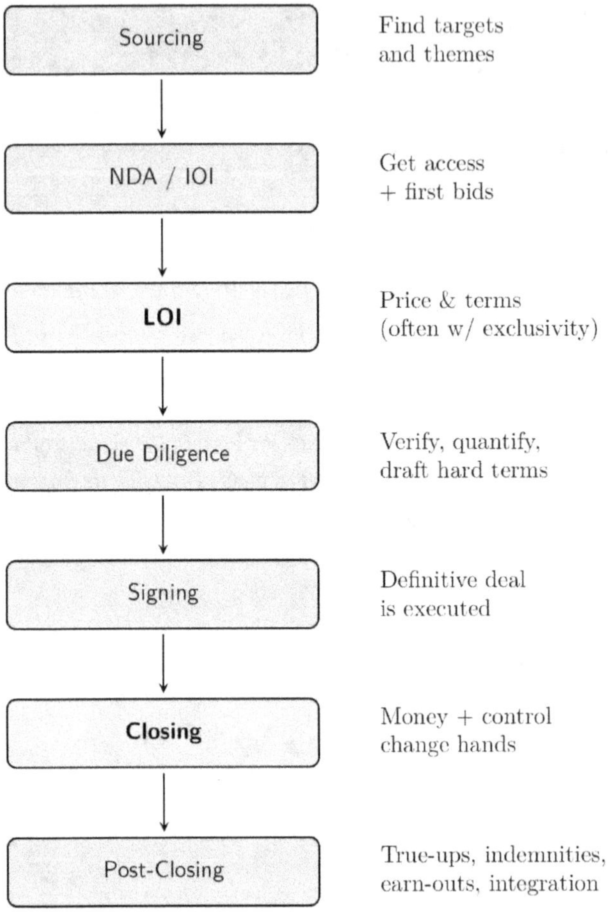

A deal is not one event; it's a controlled transfer of risk from seller to buyer, traded in stages for information, time, and (eventually) a signed promise. Early on, everybody speaks in probabilities: "We *think* the customer contracts will renew," "We *believe* margins normalize," "We *expect* the plant has another decade in it." By the end, those soft verbs get replaced with hard ones: *represents, warrants,*

shall, indemnify. The lifecycle is the story of how guesses become obligations.

The cadence is famously "hurry up and wait" because M&A has two clocks ticking at once. One is the *process clock*—deadlines, bidders, management meetings, draft turns. The other is the *gating clock*—the items that cannot be rushed without breaking something: lender underwriting, third-party consents, regulatory waiting periods, financial statement readiness, and the simple fact that diligence answers only arrive as fast as people can find documents and tell the truth. The art is keeping the process clock loud enough to maintain leverage, without pretending the gating clock doesn't exist.

Think of the deal lifecycle as three distinct emotional states with very different physics. Before the LOI, the seller is selling *possibility*. After the LOI, the buyer is hunting *proof*. Between signing and closing, lawyers and lenders become the air-traffic controllers, and the question turns from "Do we like this deal?" to "Can we land it safely?" Each stage has its own deliverables, its own traps, and its own moments where negotiating power quietly changes hands.

Phase 1: Sourcing (where the best deals are often decided)

Sourcing sounds like a polite word for shopping. In reality, it's where strategy gets translated into a list of names, and where most value is either created or quietly lost. A buyer with a clear acquisition thesis (industry, geography, product line, customer type, margin profile) tends to see opportunities earlier and move faster. A buyer without one tends to confuse motion for progress: a thousand calls, fifty first meetings, and no coherent reason any of them should become a transaction.

On the sell side, sourcing is mostly invisible because it often happens *before* a formal sale process exists. A founder mentions in passing that she's tired. A sponsor's hold period is coming due. A division underperforms quarter after quarter, and the board's patience dries up. The seller may not be "for sale" in the way a house has a sign out front, but the market can smell when a company is becoming available. In many industries, the earliest inbound call is the real beginning of the deal, even if everyone keeps pretending it's just "a conversation."

The practical output of sourcing is not a price. It's a *target list* and a *narrative*. The narrative matters because M&A is a persuasion business long before it's a documentation business. "We can expand your product into our distribution" is a story. "We can cut $3 million of overhead" is a story. "We need scale to survive procurement" is a story. Buyers that can articulate a credible story tend to win processes even when they are not the absolute highest bidder, because sellers crave certainty and a clean path to close.

Sourcing also determines who has leverage later. If a seller has multiple credible buyers circling, the LOI stage becomes an auction with deadlines. If only one buyer is truly motivated, the LOI stage becomes a negotiation with long pauses and endless "one more question." The documents look the same in both cases. The power dynamics do not.

Phase 2: NDA / IOI (permission to see the real company)

The NDA is the bouncer at the club: no signature, no access. But in practice, it's more than confidentiality language. It sets expectations for how information can be used, who can see it, and what happens if the buyer is compelled to disclose (for example, by subpoena). Sellers often add behavioral rules too—sometimes even limitations on contacting employees, customers, or suppliers without permission. The buyer may treat these as boilerplate. The seller often doesn't. In a dispute, "boilerplate" becomes "binding."

Once the NDA is in place, the seller starts revealing the company in layers. The first layer is designed to be read quickly. The goal is not completeness; it's to help serious buyers form a point of view without drowning them. The seller wants enough detail to attract the right buyers and enough mystery to preserve negotiating posture. Buyers want enough detail to avoid wasting time. This tension is why the early phase often feels simultaneously informative and frustrating.

The IOI (Indication of Interest) is the buyer's first attempt to put shape around intention without putting skin in the game. It's typically non-binding and deliberately flexible: a valuation range, a proposed structure, and a list of conditions (financing, confirmatory diligence, management meetings). An IOI is like saying, "If the house looks the way the photos suggest, I might offer something like this." It signals seriousness, but it also preserves escape routes.

Sellers use IOIs to sort the field. They're looking for three things that rarely live together in one bidder: (1) price, (2) certainty, and (3) cultural/strategic fit. The trick is that the highest initial number is often the least reliable number. Strong sellers learn to ask a quiet follow-up question that cuts through the fog: *What would have to be true for you to pay the top of your range?* A thoughtful answer is worth more than a flashy headline.

A subtle shift happens right here: information begins to have a price. Every additional schedule, customer cohort analysis, churn table, backlog roll-forward, or capex history helps the buyer underwrite the deal. But it also reduces the seller's ability to rely on ambiguity later. Sellers want buyers to become convinced; they don't want to become exposed. Managing that trade is one of the first real skills of a sale process.

Phase 3: LOI (the deal gets a backbone)

The LOI (Letter of Intent) is the milestone that turns a courtship into an engagement. It is usually still mostly non-binding on the big economic promises, but it begins to act like a spine: it supports the rest of the process, and it signals to everyone involved that there is now a default path forward. This is why LOIs are negotiated with more intensity than their "non-binding" label suggests. You can change an LOI later, but only by paying for it in trust, time, and sometimes money.

What makes the LOI powerful is not that it *closes* the deal; it *organizes* the deal. It anchors price (or at least a framework for price), structure (asset vs. stock, merger, rollover equity), timing, exclusivity, and the major conditions. It also becomes the reference point for every future argument: "That's not what we agreed in the LOI." Even when the LOI is technically non-binding, the social cost of deviating from it is very real.

Exclusivity is the LOI's sharp edge. When a seller agrees to a no-shop or exclusivity period, they are trading competitive tension for a single buyer's focused effort. In a strong market, sellers hesitate to grant this too early. In a weak market, buyers demand it as the price of investing real diligence dollars and executive time. Exclusivity is also where leverage often shifts. Once the seller closes the door to other bidders, the buyer has more room to ask hard questions, demand revisions, and renegotiate. That's not inherently bad; it's the natural consequence of choosing a single counterparty. But it is why sophisticated sellers treat exclusivity like a scarce resource.

The LOI stage is also where "headline price" starts getting its fine print. Buyers begin to specify concepts like cash-free, debt-free, working capital targets, escrow amounts, earn-outs, and seller notes. None of those are mere mechanics. Each is a lever that changes who bears risk and when. A seller who celebrates a number without reading the mechanics is celebrating a movie poster.

A clean LOI is not necessarily the shortest LOI; it's the LOI that prevents surprise. Surprises don't just delay deals. They change the emotional tone. A buyer who feels surprised becomes defensive and begins to draft for every hypothetical. A seller who feels surprised becomes suspicious and begins to assume bad faith. The LOI's real job is to keep both parties inside a narrow corridor where they can still like each other while negotiating like professionals.

Phase 4: Due Diligence (where optimism meets receipts)

Due diligence is where the buyer tries to prove two things: (1) the business is what it appears to be, and (2) the buyer understands it well enough to own it without waking up in a cold sweat six months later. It's called "diligence" because it is fundamentally about care: did you look where a careful buyer should look? But it's also a negotiation tool. The buyer is collecting facts that will become either comfort or leverage.

Diligence expands in concentric circles. First comes the core: financial statements, tax returns, customer concentration, key contracts, HR and benefits, litigation, IP, compliance, and a tour of operations. Then come the questions that reveal how the business behaves under pressure: pricing discipline, churn by cohort, backlog quality, warranty claims, supplier fragility, seasonality, capex cycles, and what happens when the founder takes a week off. Good diligence is less like a checklist and more like a detective story with a clear theory of the case.

A common misconception is that diligence is primarily about finding fraud. Most of the time, it's about finding *misalignment* between how the seller describes the business and how the business actually produces cash. A company can be honest and still be misunderstood. Revenue recognition policies can be consistent and still be aggressive. Customer relationships can be strong and still be concentrated. The buyer's job is to convert these realities into deal terms: price adjustments, escrow, specific indemnities, covenants, closing conditions, or even a decision to walk.

The seller's job during diligence is not to "win" every question. It's to keep momentum without giving away control. That requires a disciplined information flow: a clean data room, a single Q&A channel, a cadence for responses, and internal alignment on what the truth is. Sellers often underestimate how quickly buyers notice chaos. A messy diligence process signals operational mess, even when the underlying business is sound. Conversely, a well-run diligence process can make a merely good business feel like a safe one.

The workstreams run in parallel, which is why diligence can feel like a blender. While the finance team is debating normalization adjustments, legal is reading contracts for change-of-control clauses, HR is fielding benefit plan questions, IT is answering cybersecurity questionnaires, and the CEO is being asked to predict the future with uncomfortable specificity. The buyer's team is doing the same, plus building an integration plan and a financing package. This is the stage where the deal can either become a coordinated machine or a calendar war.

One of the most important psychological shifts happens here: the buyer begins to picture owning the company, and that changes behavior. Before diligence, buyers speak in hypotheticals. During diligence, they start asking operational questions that betray intent: "Who approves vendor onboarding?" "How quickly can we change pricing?" "What does the Monday morning meeting look like?" These are the questions of someone imagining responsibility. Sellers should notice this; it often signals the buyer is moving toward commitment, even if the spreadsheets are still being argued.

Phase 5: Signing (the moment the deal becomes a legal organism)

"Signing" is not "closing." Signing is when the definitive agreement is executed—the purchase agreement becomes real, and the parties become bound to a path that ends either in closing or in a documented breakup. This is where the deal stops being a set of intentions and starts being a legal organism with organs: representations and warranties (what's true), covenants (what you'll do), closing conditions (what must happen), and remedies (what happens if it doesn't).

Drafting is where the deal's philosophy shows up. A buyer who expects turbulence drafts a purchase agreement like a fortress: tight definitions, broad reps, long survival periods, large escrows, and aggressive termination rights. A buyer who trusts the seller drafts a document that assumes normal life and focuses on the big risks. Neither approach is morally superior; each reflects risk appetite, industry norms, and negotiating leverage. But sellers should recognize what the draft is telling them: not just what the buyer *wants*, but how the buyer *sees* the world.

The most consequential battles at signing are rarely about the obvious price. They're about who owns ambiguity. Definitions become destiny. What counts as "debt" in a cash-free, debt-free deal? What is "working capital" and what accounting policies govern it? Which liabilities stay with the seller, which transfer, and which are shared through indemnity? A single defined term can swing millions of dollars. This is why experienced deal teams obsess over the definition section even when it feels pedantic. Pedantry is just precision with bad branding.

Signing also introduces the deal's escape hatches. Termination provisions define when either side can walk away and what happens next. Specific performance language may determine whether a party can be forced to close. Reverse termination fees, if present, price the buyer's right to fail to close (often tied to financing risk). These terms don't just affect litigation posture; they affect behavior. A buyer facing meaningful consequences for walking is more likely to solve problems rather than manufacture them.

If there is a gap between signing and closing (common in regulated transactions or those requiring third-party consents), the deal enters an interim period with its own rules. The seller typically agrees to operate "in the ordinary course" and not do anything that would change the business's risk profile. The buyer often gains consent rights over major actions: new debt, major capex, key hires or terminations, unusual contracts. This is the strange phase where the seller still owns the company but no longer has full freedom, and the buyer has influence without yet having responsibility. It is fertile ground for friction unless expectations are explicit and communication is frequent.

Phase 6: Closing (money, control, and a thousand small signatures)

Closing is where the transaction stops being a narrative and becomes logistics. Funds move. Shares transfer (or assets are conveyed). Board resignations happen. Officer certificates are signed. Legal opinions may be delivered. Lenders wire into escrow accounts with their own checklists. If you want a realistic picture of closing, imagine an orchestra tuning up: dozens of parties doing small actions that must land in the right order, at the right time, with the right documents, often across multiple time zones.

There's a reason experienced deal people obsess over the closing checklist. It's not because they love paperwork. It's because closings fail in boring ways: a missing payoff letter, a consent that didn't arrive, a signature block with the wrong legal name, a wire instruction that wasn't verified, a lien release that wasn't properly executed. The economic deal can be perfect and still not close if the plumbing leaks.

The purchase price mechanics often show up in the closing flow like hidden gears. Is there an escrow or holdback? That changes where the funds go and who controls them. Is there a working capital adjustment with a post-close true-up? That changes how much cash actually lands with the seller on day one. Is there assumed debt that must be refinanced or paid off at closing? That changes the wiring diagram and the timing. The headline number might be printed in the press release, but the closing statement is where the truth lives.

Control shifts at closing, but responsibility does not always shift cleanly. Many deals come with transition services agreements (TSAs), short-term arrangements where the seller continues providing payroll, IT, accounting, or facilities support while the buyer stands up replacements. Even without a formal TSA, there is usually a human TSA: the seller's team still knows where everything is, and the buyer still needs them to answer questions. A smooth close treats that reality with respect rather than pretending the ink magically transfers institutional knowledge.

Post-closing: when the deal starts keeping score

The first surprise for many first-time sellers is that closing is not the end of the math. It's the start of the accounting. If the deal includes working capital adjustments, earn-outs, escrow releases, indemnity claims, or purchase price allocation true-ups, there are future moments when money can move again. Those moments can be friendly, mechanical check-ins—or they can become slow-motion disputes that sour what was once a celebratory transaction.

Working capital true-ups are the classic example of a "small" term that can become a large fight. The concept is simple: the buyer expects the business to be delivered with a normal level of operating liquidity (current assets minus current liabilities, defined carefully). If closing working capital is above or below a negotiated target, the price adjusts. The complexity comes from definitions and accounting policies. Two competent accountants can produce different answers if the agreement gives them room to interpret. The seller who treated working capital as a footnote sometimes learns, post-close, that footnotes can be expensive.

Indemnification is another post-close scoreboard. Reps and warranties are promises about the business as of signing (and sometimes as of closing). If a promise turns out to be untrue, the buyer may have a claim. Sellers often view indemnity as an annoying tail. Buyers view it as the seatbelt: you hope you never need it, but you don't drive without it. The practical question is how much of the purchase price remains at risk (escrow/holdback), for how long (survival), and with what friction (baskets, caps, procedures). This is where the deal's risk allocation becomes real money.

Earn-outs, when used, extend the relationship in the most delicate way possible: by tying future payments to future performance. Done well, they bridge valuation gaps. Done poorly, they create incentives for both sides to interpret the business differently. Sellers may feel the buyer is "managing down" results. Buyers may feel the seller is "gaming" metrics or clinging to old operating habits. The smartest earn-outs are designed with two questions in mind: *Can it be measured unambiguously?* and *Will the buyer run the business differently post-close?* If the answer to either is "yes," simplicity becomes a virtue, not a concession.

Then there's integration—the part of the deal that determines whether the acquisition becomes a value-creating move or an expensive lesson. Integration is not only systems and org charts. It's the buyer's first real test of whether they understood what they bought. Culture shows up immediately in small moments: how decisions are made, how fast approvals happen, what gets measured, how conflict is handled, how customers are treated when something goes wrong. Many deals fail to create value not because the diligence missed a fraud, but because integration underestimated human physics.

The lifecycle has a repeating pattern: each phase reduces uncertainty, but each reduction costs something. Sourcing costs time and attention. NDA/IOI costs confidentiality risk and competitive exposure. LOI costs optionality (especially with exclusivity). Diligence costs money, bandwidth, and sometimes ego. Signing costs flexibility. Closing costs precision. Post-close costs patience. The dealmaker's job is to pay those costs in the right order, at the right scale, without overpaying early for certainty that could have been purchased later.

A final, underappreciated truth: most "deal problems" are lifecycle problems. A buyer who tries to negotiate purchase agreement terms as if they were still at IOI stage creates whiplash. A seller who treats diligence like a marketing exercise rather than a verification exercise invites repricing. A team that rushes signing without a clean closing plan creates avoidable closing drama. Understanding the lifecycle doesn't just make you fluent in process. It makes you better at timing—and timing, in M&A, is often the difference between a clean win and a costly almost.

1.2 The Deal Team: Roles and Advisors

Deals don't collapse because someone can't do algebra. They collapse because the wrong people are in the room, the right people are in the room too late, or everyone is in the room but nobody owns the decision. M&A is a team sport with a weird scoreboard: half the points are saved by avoiding mistakes that never make headlines, and the other half are scored by moving fast without breaking trust.

The cast is larger than most first-timers expect, and the roles are more distinct than their job titles suggest. A CFO can behave like a project manager, a prosecutor, or a therapist depending on the day. A lawyer can be a tactician or a traffic cop. A banker can be a strategist or a metronome. The quickest way to understand the deal team is to sort people into two groups: *Principals* (the ones whose names are effectively on the check) and *Advisors* (the ones paid to keep that check from becoming a regret).

Principals: the people who own the risk

On the buy side, the Principal is whoever is truly accountable for the acquisition working in the real world, not just closing on paper. That might be a corporate development (Corp Dev) leader acting for a strategic buyer, a private equity partner acting for a fund, or an owner-operator buying a business with a lender. Titles can mislead. The tell is simple: if the deal goes sideways, whose Monday morning gets ruined?

Corp Dev is often the buyer's quarterback, but the best Corp Dev teams are allergic to playing hero. They can model and negotiate, but their real value is orchestration: pulling in finance, legal, tax, HR, IT, and the business unit at the right moments, then translating everyone's concerns into terms the deal can actually absorb. Corp Dev is also the institutional memory. They remember what happened the last time a customer consent delayed closing, or what a specific rep-and-warranty gap cost in the last acquisition, and that memory quietly shapes today's risk tolerance.

For a strategic buyer, the business unit sponsor (the GM, division president, or product leader who will inherit the asset) is a Principal whether they like it or not. They may not lead diligence calls, but they will live with the integration, the customer churn, the cultural friction, and the operational surprises. When that sponsor is absent, a deal can become a corporate exercise rather than an operating decision. The result is a purchase agreement that closes cleanly and a business that never quite clicks.

On the sell side, the Principal might be a founder, a family, a board, a private equity sponsor, or a corporate parent divesting a division. Here, too, the name on the signature page isn't always the true Principal. Sometimes the CEO signs, but the board decides. Sometimes a sponsor decides, but management carries the relationship risk with employees and customers. The best sale processes treat these as separate audiences with different anxieties: the board worries about certainty and price; management worries about disruption and dignity; owners worry about taxes and net proceeds; everyone worries about reputational fallout.

A special kind of Principal appears in many deals: the *management team rolling equity*. Rollover equity simply means the seller (often management) reinvests part of their proceeds into the new ownership structure instead of taking all cash at closing. The minute that happens, management becomes both seller and buyer—selling today's business while buying tomorrow's future. That dual identity changes the room temperature. Negotiations become less about "getting every dollar now" and more about "not poisoning the business we still have to run."

Then there's the Principal everyone forgets until the last week: the lender. In leveraged acquisitions, the lender doesn't just supply funds; it supplies permission. Credit committees have their own diligence expectations, their own documentation checklists, and their own timeline. If a buyer treats financing as "already handled," the lender has a way of reintroducing itself at precisely the worst possible time, usually by asking for one more report that takes two weeks to produce.

The core deal office: who keeps the train on the tracks

Every deal has a de facto *deal office*, even if nobody calls it that. It's the small group that sets the cadence, runs the weekly calls, tracks open items, and decides what gets escalated. On smaller deals, this might be just the CEO and CFO on one side, and a Corp Dev lead on the other. On bigger deals, it's a structured project with workstreams, issue logs, and a closing checklist that reads like a novella.

The most effective person in the deal office is often not the most senior. It's the one with the discipline to say, politely and repeatedly, "Who owns this, and when will it be done?" M&A rewards charm, but it runs on follow-through. A brilliant negotiator can't save a deal that misses its own internal deadlines.

Expect the deal office to spend an unreasonable amount of time on what looks like administrative trivia: access lists, document versions, the question log, draft redlines, and who is allowed to talk to which customer. That "trivia" is actually risk management. Deals leak through tiny cracks: an outdated financial schedule, an employee answering a diligence question off-the-cuff, a buyer team member saving a confidential file in the wrong place. The deal office is there to reduce the number of unforced errors.

Investment bankers: market-makers, psychologists, and conductors

If lawyers are guardians of liability, bankers are guardians of *process*. A good investment banker is not a spreadsheet in a suit; they are the person who can keep multiple parties moving forward without letting any one party feel cornered. They run the tempo of a sale process: when IOIs are due, when management presentations happen, when bids sharpen, when exclusivity is granted. Timing is their product.

On a sell-side process, bankers create competitive tension and protect management's bandwidth. A founder selling their company doesn't want to spend three months answering the same five questions from ten different buyers. Bankers standardize the early-stage story, manage the flow of information, and filter out tourists. They also deliver a social service that is hard to price: they absorb bruised feelings. When a buyer pushes too hard on a term, the banker can be the shock absorber so management doesn't have to carry the negativity.

Bankers also play the role of *reality translator*. Sellers say, "We deserve 12x because our product is unique." Buyers say, "We'll pay 7x because churn is high." A banker's job is to convert those positions into a zone where a deal can exist, using comparables, precedent transactions, and an understanding of how different buyers underwrite risk. The banker is often the first person to say out loud what everyone privately suspects: "This number is aspirational," or "This buyer is serious, but they need structure to get there."

On buy-side mandates (common in some strategic acquisitions and sponsor add-ons), bankers can help find targets, open doors, and shape outreach. But their most valuable function is often *triage*. They can quickly tell a buyer which opportunities are real and which are mirages, and they can bring pattern recognition from hundreds of conversations that a single buyer simply doesn't have.

The banker's incentives deserve respect, not cynicism. Yes, success fees exist. But good bankers know their franchise depends on more than a closing dinner. Sophisticated clients remember who pushed for a fragile deal that later imploded, and they remember who told an uncomfortable truth early enough to change course.

Legal counsel: the guardians of liability (and the engineers of precision)

M&A lawyers are often described as deal-killers by people who haven't needed them yet. Their real job is to prevent a celebratory signing from becoming a multi-year argument about what everyone "meant." They do this by turning business intent into enforceable language, and by anticipating how a dispute would look to a third party who doesn't care about anyone's feelings: a judge, an arbitrator, or an insurer.

There are usually two flavors of legal work. The first is *architecture*: structuring the transaction, choosing the form of agreement (stock purchase, asset purchase, merger), and mapping the closing conditions and deliverables. The second is *precision*: writing and negotiating definitions, reps and warranties, covenants, indemnification frameworks, and the procedures for post-closing adjustments and claims. Architecture gets the deal built. Precision keeps the deal from collapsing under its own ambiguity.

Good lawyers ask questions that sound annoying but are actually clarifying. "What exactly counts as debt?" "Which accounting policies define working capital?" "Who decides whether a claim is valid?" "What happens if the consent doesn't arrive?" These questions are not academic. They are the difference between a clean close and a dispute that eats management's attention for a year.

Legal counsel also manages *process risk*. They coordinate signatures, corporate approvals, charter documents, board minutes, payoff letters, lien releases, and the closing checklist. Many of the most dangerous deal failures are procedural: the buyer wires funds but a lien isn't released; the seller signs but doesn't have authority; a required third-party consent is missed. These are not glamorous problems, but they can be existential.

Specialists often appear under the legal umbrella: employment counsel for benefit plans and executive arrangements, IP counsel for patents and software licensing, real estate counsel for leases and title matters, and regulatory counsel for industry-specific rules. The presence of specialists is usually a good sign: it means the lead counsel knows what they *don't* know.

Accountants and financial diligence providers: the people who argue with the numbers so you don't have to later

Financial diligence teams (often from accounting firms or specialized advisory shops) are the ones who take the seller's financial story and stress-test it under a buyer's definition of reality. They build a bridge between accounting statements and deal economics: quality of earnings (QoE), net working capital analysis, net debt, and sometimes quality of revenue. The product is not a single number; it's a defensible understanding of *how the business produces cash* and *where the risks hide.*

A QoE report is frequently misunderstood as a forensic audit. It's not designed to opine on whether the financial statements are fairly presented under GAAP in the way an audit does. It's designed to help a buyer answer a more practical question: "What earnings level should I pay for if I own this business next year?" That means identifying non-recurring items, one-time add-backs, owner perks, unusual accrual swings, and revenue/cost timing issues that can make EBITDA look healthier (or weaker) than it will be under new ownership.

Working capital is where these teams earn their keep. Working capital adjustments are common, and disputes are common because the concept is simple but the definitions are not. Does "working capital" include deferred revenue? What about accrued bonuses? How are inventory reserves treated? Which accounting policies govern the closing statement? A financial diligence team can spot the landmines early and help draft a working capital definition that matches the business's real operating cycle rather than an abstract textbook formula.

These teams also function as a credibility engine. When a buyer's investment committee asks, "Why are we comfortable with this margin profile?" the answer "management said so" is not enough. A third-party diligence report gives the buyer internal permission to proceed and gives the seller a cleaner path to close by reducing last-minute skepticism.

Tax advisors: the quiet architects of what people actually take home

Tax is where deals go from "good" to "why didn't anyone tell me that?" A purchase price that looks attractive on a headline basis can produce very different after-tax outcomes depending on structure, allocation, and timing. Tax advisors are brought in to design the transaction so it works economically *net of taxes*, and to prevent a future unpleasant letter from a tax authority.

On the buy side, tax advisors care about basis and future deductions: how much of the purchase price can be amortized, how quickly, and in what jurisdictions. On the sell side, tax advisors care about characterization: capital gain vs ordinary income, the treatment of rollover equity, and the tax impact of escrows, earn-outs, and working capital true-ups. The same term can feel minor to a deal team and enormous to a tax advisor because tax outcomes can swing by millions with a small structural change.

Tax advisors also help with cross-border complexity, state and local tax exposures, transfer taxes, and the treatment of transaction costs. They are often the ones who insist on seemingly fussy language about "purchase price adjustments" versus "indemnity payments" because those labels can change tax reporting.

The best tax advisors show up early enough to influence structure rather than late enough to merely describe the damage. A last-minute tax scramble usually means the commercial team made assumptions that were never guaranteed.

Valuation and M&A specialists: when ``reasonable" needs to be defensible

Valuation specialists appear when someone needs an answer that will be scrutinized by auditors, regulators, or future litigants. Purchase price allocation for financial reporting (allocation to tangible assets, identifiable intangibles, and goodwill) is one common driver. Another is fairness opinions in certain transactions. Another is disputes: earn-out calculations, working capital true-ups, or indemnity claims that turn into accounting debates.

Even when a deal is friendly, a valuation specialist can be a useful pressure test. They can identify whether the deal's assumptions are internally consistent: growth rates, margin expansion, customer retention, capital intensity. They won't run the business, but they can tell you when the story violates arithmetic.

Insurance brokers and reps & warranties insurance: shifting risk to a third party

Reps and warranties insurance (RWI) has become a common tool in many private-company deals, particularly when sellers want a cleaner exit and buyers want protection without a massive escrow. An insurance broker (and the insurer's underwriting team) effectively becomes another diligence audience with its own requirements and timelines.

RWI changes the negotiation. Sellers may push for lower escrows and tighter indemnity caps because the buyer can look to the policy. Buyers may accept a different indemnity package because they value the certainty of an insurer-backed recovery. But the policy is not magic. It has exclusions, retention (a deductible-like amount), and underwriting-driven diligence expectations. If the deal team treats RWI as a checkbox, it can create a rude surprise when the insurer excludes the exact risk everyone was worried about.

HR, IT, and operational specialists: the deal's hidden risk surface

HR diligence is rarely about whether employees exist. It's about whether the workforce can survive the transaction without bleeding value. Who are the key people? Are there change-of-control payments? Are there misclassification risks (contractor vs employee)? Are benefits plans properly administered? What's the tone of employee relations? A buyer can buy revenue, but it cannot buy morale at the closing table.

IT and cybersecurity diligence has moved from "nice to have" to "board-level." Data breaches, weak access controls, unsupported systems, and fragile integrations can turn a promising acquisition into a costly remediation project. IT specialists will ask questions that sound technical and feel intrusive because they are trying to map operational continuity risk: can the business keep billing, shipping, and serving customers on day one under new ownership?

Operational specialists vary by industry: environmental consultants for industrial sites, engineers for manufacturing processes, regulatory experts for healthcare or financial services, and supply chain specialists for businesses with single-source dependencies. These experts often deliver a single sentence that changes the deal: "That facility expansion will require a permit timeline you're not modeling," or "Those customer contracts are assignable only with consent." One sentence can be worth more than a thousand pages of general diligence.

The data room administrator: the most underrated role in the deal

Somebody has to run the virtual data room (VDR) like it matters, because it does. The VDR administrator controls permissions, versioning, watermarks, uploads, and the audit trail of who viewed what and when. That sounds like clerical work until you realize the VDR is where diligence either accelerates smoothly or dies of friction.

A disciplined VDR admin makes a company feel investable. Documents are labeled consistently, key agreements are easy to find, financials tie out, and questions get answered in a predictable flow. A chaotic VDR admin does the opposite: it signals disorganization, increases buyer skepticism, and invites a thousand repetitive requests that grind management into dust.

The audit trail is also strategic. When a dispute arises later about disclosure, the ability to show that a document was uploaded on a certain date and accessed by certain users can shift outcomes. The VDR is not just a filing cabinet; it's part of the deal's memory.

How the team evolves as leverage shifts

The deal team doesn't stay constant. Early on, it's sales and strategy heavy: bankers, Corp Dev, and the CEO. After an LOI, the roster thickens: diligence providers, specialists, and more lawyers. Closer to signing and closing, the team becomes document-heavy and checklist-driven. Post-close, accountants and integration leads often become the center of gravity. Each phase changes which skills matter most.

This is why experienced dealmakers are careful about when to introduce which voices. Bring in lawyers too early and you can suffocate commercial momentum with hypotheticals. Bring them in too late and you discover that "standard" terms aren't standard for your deal. Bring in tax late and you can't fix structure without reopening economics. Bring in IT late and you can close into a systems mess that costs more than the purchase price adjustment you negotiated so hard.

The cleanest deals usually have one human pattern: a small number of decision-makers who stay engaged, and a larger circle of experts who are empowered to be precise without being political. Precision is not the enemy of speed. Politics is.

A practical way to read the room

When you're trying to understand who really matters in a given deal, don't start with the org chart. Start with three questions.

First: who can say "no" without asking permission? That's a Principal, regardless of title.

Second: who owns the model that drives the buyer's price and terms? That person has leverage, because numbers are the language deals use to justify decisions.

Third: who writes the definitions that will govern money movement after closing (debt, cash, working capital, earn-out metrics)? That person is shaping reality. In M&A, the pen can be as powerful as the checkbook.

A good deal team doesn't just fill seats. It creates an environment where uncomfortable truths surface early, where specialists are used to clarify rather than to posture, and where the Principals stay close enough to make real-time decisions. The transaction may be negotiated in documents, but it's won or lost in coordination.

1.3 Structuring the Transaction

Structure is the deal's skeleton. Price is what everyone argues about out loud; structure is what decides who inherits the messes that weren't argued about at all. Two transactions can share the same headline number and feel like completely different worlds at closing: one where the buyer steps into an existing legal entity like putting on a suit, pockets and all; another where the buyer cherry-picks only what it wants and leaves the rest behind like an abandoned storage unit.

The simplest way to think about structure is to ask a blunt question: *What, exactly, is the buyer buying?* Not the business in the philosophical sense—the legal thing. Is it the equity of a company (its stock or membership interests)? Is it a list of assets (contracts, equipment, inventory, IP) assembled into a functioning operation? Or is it a statutory combination where one entity merges into another under corporate law? Those are the three foundational shapes. Everything else—tax elections, earn-outs, escrows, rollover equity—gets bolted onto one of them.

A second question matters just as much, and it's less romantic: *What liabilities come along for the ride?* "Liability" here is not only lawsuits. It's unpaid payroll taxes, messy employee classifications, a lease you forgot renews automatically, warranties that will be called next year, an underfunded pension plan, a cyber incident nobody wants to talk about, or a customer contract that quietly forbids assignment. Structure is how you decide whether those risks stay with the seller, move to the buyer, or get shared through contractual protection.

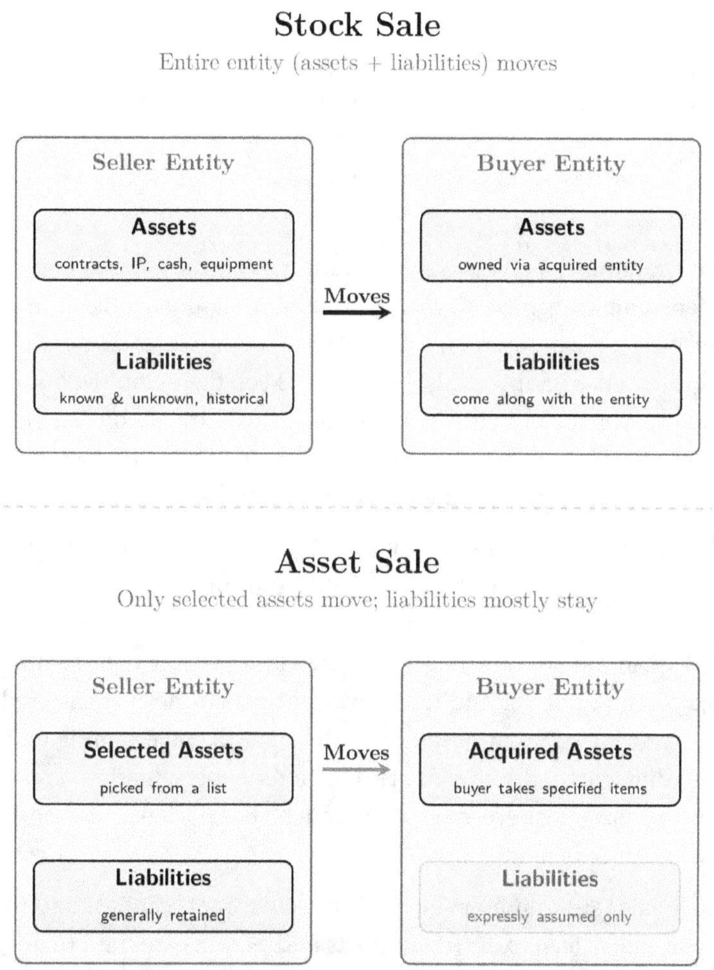

The diagram is intentionally blunt because the difference really is that blunt: in a stock sale the buyer acquires the *entity*; in an asset sale the buyer acquires a *bundle*.

Stock sale: buying the company, history included

In a stock sale (or membership interest purchase for an LLC), the buyer purchases the equity of the target entity. The entity continues to exist; it still owns the same assets and owes the same obligations. The buyer has simply replaced the shareholder at the top of the chain.

That continuity is the stock sale's superpower. Customer contracts often remain in place without needing assignment. Licenses and permits may continue uninterrupted (though sometimes they still require notice or approval). Employees keep their employer, at least legally, which can make benefits and payroll transitions smoother. Operationally, a stock deal can feel like changing the lock on the front door rather than moving the whole house.

Continuity also means inheritance. When you buy the entity, you inherit its past—everything the entity did, promised, failed to do, and might be accused of doing. Some of those liabilities are visible and quantifiable. Others are contingent: a lawsuit not yet filed, a tax position not yet audited, a warranty claim not yet asserted, a regulatory issue that hasn't been discovered. Buyers mitigate this with diligence and contractual protections (representations and warranties, indemnities, escrows, insurance), but the basic fact remains: the entity is the entity.

Stock sales are common when (a) the seller wants capital gains treatment and a clean exit, (b) contracts and permits are hard to assign, (c) the business is difficult to carve into assets cleanly, or (d) there's meaningful value in the legal continuity itself—think long-term customer agreements, regulated operations, or businesses whose IP and contracts are heavily intertwined.

The purchase agreement in a stock sale tends to spend serious time on *reps and warranties* about historical matters: taxes paid, compliance with laws, accuracy of financials, absence of undisclosed liabilities. That's not paranoia; it's the logical response to buying something with a memory. The reps and warranties are the buyer's way of converting "we believe" into "you promise."

Asset sale: buying the operation as a curated list

In an asset sale, the buyer purchases specified assets and assumes specified liabilities—usually the ones needed to operate the business day-to-day. Everything not listed stays with the seller by default (or more precisely: it stays with whatever entity is left behind).

Asset deals appeal to buyers who want control over what they're taking on. The purchase agreement often reads like a packing list: *Acquired Assets* and *Excluded Assets*; *Assumed Liabilities* and *Excluded Liabilities*. That inventory mentality is the asset deal's defining feature. It's also why asset deals can feel document-heavy. When you buy the entity, you don't need to list each wrench in the toolbox. When you buy assets, you do—or you accept the risk that you didn't.

This structure is common when the target has messy legacy liabilities, when the buyer wants a specific division or product line rather than the whole company, or when the seller is in a context like a restructuring where asset transfers are the cleanest way to move value. It's also common in smaller "Main Street" acquisitions where buyers (often with bank financing) want a clear picture of what they own and what they don't.

But the "liabilities stay behind" story is not as absolute as people hope. There are legal doctrines in many jurisdictions that can impose successor liability in specific circumstances (for example, certain tax liabilities, environmental obligations, or employment-related claims), and there are practical realities that create pressure to assume more than expected. If you want the customer contract, you may have to assume the obligations attached to it. If you want the workforce, you may need to adopt certain benefit obligations or honor certain accrued liabilities to keep people from leaving.

Asset deals also collide head-on with *consents*. Many contracts prohibit assignment without the counterparty's approval. In a stock sale, there may be no assignment at all (though change-of-control clauses can still trigger consent requirements). In an asset deal, assignment is often the point, and consents can become the timeline. One stubborn landlord or one major customer can turn a clean economic deal into a prolonged negotiation about leverage that has nothing to do with price.

A useful mental model: asset deals buy *things*; stock deals buy *relationships*. The more a business depends on relationships embedded in contracts, permits, and people, the more structure becomes a question of portability.

The tax gravity of asset deals: allocation and Form 8594

Taxes don't politely wait for closing; they shape structure from the beginning because they shape what the buyer gets to deduct and what the seller gets to keep.

Asset acquisitions typically require the purchase price to be allocated among different classes of assets for U.S. tax purposes under Internal Revenue Code §1060. The buyer and seller generally report this allocation on IRS Form 8594 (Asset Acquisition Statement). This is not mere paperwork. Allocation determines the buyer's tax basis in each asset—essentially the starting point for future depreciation and amortization—and it affects how the seller's gain is characterized.

Why does that matter in plain English? Because a dollar allocated to equipment is not the same as a dollar allocated to goodwill. Different asset classes get written off over different schedules. Buyers generally like allocations that create faster deductions; sellers often prefer allocations that yield more favorable tax outcomes (for example, capital gain versus ordinary income, depending on the asset and the seller's facts). The allocation is a negotiation, and it's one reason asset deals can become surprisingly technical late in the process.

One twist that catches people off guard: if consideration changes after closing—because of an earn-out, a working capital true-up, or an indemnity adjustment—allocation may need to be revisited. A "final price" that moves post-close can have tax reporting consequences, which is why tax advisors stay involved even when everyone else is mentally finished.

Mergers: the statutory machinery that can make hard things easier

A merger is less a third economic idea and more a legal mechanism for combining entities under corporate law. In a statutory merger, one company merges into another, and by operation of law, assets and liabilities transfer. The buyer ends up owning the surviving entity (directly or indirectly), and the disappearing entity ceases to exist.

Mergers are often used when the target has many shareholders (especially in public company transactions), when state corporate law offers a clean pathway for approvals, or when the transaction needs a particular legal effect—like automatically transferring assets without individually assigning each one. Think of a merger as a legal conveyor belt: instead of picking up each asset and moving it by hand, the law treats the whole entity as having shifted into the survivor.

That said, merger structure does not magically delete the liability question. If anything, mergers embrace continuity. The liabilities typically come along just as they would in a stock acquisition, because the business's legal history remains attached to the surviving entity. The difference is how you accomplish the combination and what approvals you need to make it valid.

Mergers also come in variations that matter in practice: forward mergers, reverse triangular mergers (where a buyer subsidiary merges into the target, leaving the target as the survivor), and other configurations designed to achieve a particular outcome around contracts, permits, or tax treatment. The vocabulary can feel like inside baseball, but the motivation is usually simple: someone wants the transaction to count as "no assignment" for key contracts, or wants a specific tax result, or wants a cleaner shareholder approval process.

A practical comparison that actually predicts work

Structure choices aren't philosophical; they predict how painful the deal will be to execute. A practical comparison frames the comparison in the language of real friction: consents, diligence depth, tax mechanics, and post-close risk.

Dimension	*Stock Sale (Equity Purchase)*	*Asset Sale*
What transfers	The target entity (and everything it owns/owes) continues under new ownership.	Only the assets listed in the agreement; only liabilities expressly assumed (with exceptions in law/practice).
Liability profile	Buyer inherits historical liabilities of the entity; mitigated through reps/indemnities/escrow/insurance.	Buyer aims to limit liability to assumed items, but must manage successor liability risk and operational necessity.
Contract/permit portability	Often smoother, but watch change-of-control clauses and regulated approvals.	Often consent-heavy due to assignment restrictions; can drive timeline and leverage.
Tax mechanics (U.S. lens)	Often seller-friendly (capital gains for individuals), but depends on facts and elections. Buyer may not get step-up in asset basis unless special elections apply.	Purchase price allocation under IRC §1060; reporting on Form 8594; allocation affects buyer deductions and seller gain characterization.
Documentation feel	Fewer asset-by-asset schedules; heavier focus on historical	Heavier schedules: what's in/out; transfer documents;

Why buyers and sellers often prefer different structures

Structure is one of the most common places where buyer and seller preferences diverge for rational reasons.

Sellers often prefer stock deals because they feel like a clean handoff: sell the entity, distribute proceeds, and move on. For many individual sellers, stock sales can be tax-advantaged relative to asset sales, though every situation depends on facts and advisors should model it. Sellers also dislike the administrative sprawl of asset deals: the endless assignments, the consent chasing, the fear that something valuable accidentally lands in "Excluded Assets" because somebody drafted a schedule at 2 a.m.

Buyers often prefer asset deals when they worry about unknown liabilities or want a specific piece of a broader company. An asset deal can also be attractive when the buyer wants a step-up in basis for tax purposes, again depending on facts and elections. Buyers may also like the conceptual cleanliness: "We bought these things, and we assumed these obligations." That sentence is psychologically soothing—even when the real world complicates it.

A sophisticated negotiation recognizes that structure preference is not stubbornness; it's risk pricing. If a buyer insists on an asset deal, they are usually saying: "We are not willing to pay stock-deal pricing for stock-deal risk." If a seller insists on a stock deal, they are often saying: "We are not willing to accept asset-deal taxes and complexity for the same price." The solution is not to argue who is right; it's to adjust economics or protections until both sides can live with the chosen form.

How structure changes diligence---and what ``deep diligence" really means

Diligence is never just "check the financials." Structure determines where diligence spends its time.

In a stock deal, diligence leans hard into entity-level historical exposures: tax compliance, employment practices, litigation, regulatory matters, and the quality of contracts as obligations of the entity. The buyer is underwriting the *past* because the past is coming along.

In an asset deal, diligence becomes more inventory-like and operationally specific: do you actually own the assets you say you own, are they transferable, what liens sit on them, which contracts are assignable, and what consents are needed to make the asset bundle function as a business the morning after closing. The buyer is underwriting *portability*.

This is why the same target can feel "simple" under one structure and "impossibly complicated" under another. A software business with thousands of click-through customer agreements may be easier to buy as stock than as assets if assignment language is hostile. A manufacturing business with environmental history may be more comfortable as an asset deal if the buyer wants to wall off legacy exposure. Structure turns diligence into either a history exam or a logistics exam.

Execution mechanics: what actually happens differently at closing

At closing, a stock deal typically involves transferring equity, updating the cap table, delivering resignations and new officer appointments, paying off debt (if required by the deal), and ensuring the conditions are satisfied. The entity continues, so the operational transfer can be quieter than outsiders expect. Customers might not even notice on day one—sometimes by design.

At closing, an asset deal can resemble a moving day with legal labels on everything. Bills of sale, assignment and assumption agreements, IP assignments, real estate instruments, UCC filings, lien releases, contract consents, and sometimes new employment offers. If the asset bundle is large, the closing set can be thick enough to intimidate seasoned teams. None of this is inherently bad; it's simply the mechanical consequence of buying a list rather than a container.

Mergers can be mechanically elegant—file the right certificates, satisfy statutory steps, and the assets and liabilities shift by operation of law. But mergers often come with their own approval choreography (board and shareholder actions, statutory notice requirements, appraisal rights in some contexts) and, in larger deals, regulatory gating that can make the timeline feel less like a sprint and more like controlled airspace.

A quiet rule: structure should match the reason you're doing the deal

Structure is sometimes chosen because it is "market." That's a weak reason. The better reason is alignment: the structure should match the deal's true risk and the deal's true value driver.

If the value driver is continuity—keep the contracts, keep the permits, keep the workforce stable—structures that preserve the entity often win. If the risk driver is legacy exposure—tax positions, litigation, environmental history—structures that let the buyer carve out liabilities tend to dominate. If the value driver is a carve-out (a division inside a larger company), asset-style thinking becomes unavoidable even if the legal form is a stock purchase of a subsidiary, because you'll still need to define what is in and out.

One last practical truth: structure decisions are rarely final at LOI, but they become more expensive to change the longer you wait. Switching from stock to asset late in the process is not a minor edit; it rewrites consents, tax modeling, purchase price allocation, schedules, and closing deliverables. The structure is the deal's skeleton—try changing the skeleton after the house is framed and you'll discover why teams argue about it so early.

When structure is chosen well, it feels boring at the end. Boring is good. Boring means the legal form and the business reality agree with each other, and the deal team can spend its energy on the only thing that actually creates value: owning the business and making it better, rather than litigating what was bought.

1.4 Understanding Deal Economics

A "$100 million deal" is one of the most efficient lies in business. It's not usually malicious; it's just shorthand. But shorthand is dangerous in M&A because the difference between what gets announced and what gets wired can be the difference between feeling like you won and realizing you sold the roof and kept the termites.

The economics of a deal live in a few deceptively small phrases—*cash-free, debt-free, working capital target, escrow, earn-out, seller note.* Each phrase is a lever. Pull the lever and the purchase price changes form: some becomes money at closing, some becomes money later, and some becomes money only if the future behaves. The goal isn't to memorize jargon; it's to see where value can leak, where it can be deferred, and where it can quietly flip from "certain" to "conditional."

Enterprise value vs. equity value: the deal's two price tags

Most operating businesses are priced on *enterprise value* (EV). Enterprise value is the value of the business *as an operating machine,* independent of how it's financed. It's the price tag you'd put on the engine, not on the bank loan attached to it.

Equity value is what the shareholders actually own: the value left after you deal with the company's net debt position (and, in real transactions, after you account for a few other items that behave like debt). Equity value is the number that determines the equity check at closing—the portion of the deal proceeds that goes to the people selling shares, after paying off lenders and settling the balance sheet mechanics baked into the agreement.

The cleanest way to connect the two is the "enterprise-to-equity bridge":

$$\textit{Equity Value} \approx \textit{Enterprise Value} - \textit{Debt} + \textit{Cash}$$

That's the core bridge most deal teams carry in their heads. It's also why a founder can hear "$100 million" and still not receive $100 million. If the company has $30 million of debt and $5 million of cash, the bridge points toward something like $75 million of equity value before you even start talking about escrows, working capital, fees, or earn-outs.

Of course, real life is never quite that tidy. "Debt" in a deal is not always the same thing as debt on a balance sheet. It's a negotiated definition that can include items that feel like debt because they drain value at closing: unpaid taxes, transaction bonuses, certain lease obligations, deferred revenue in some deals, even unpaid professional fees if the agreement says so. This is why the definitions section in the purchase agreement can swing more money than an extra half-turn on the valuation multiple.

Cash-free, debt-free: a deceptively polite way to say ``we're going to true-up your balance sheet"

Cash-free, debt-free sounds like a principle. In practice, it's a mechanism. It means: "We're agreeing on a value for the operations. At closing, we'll adjust for how much cash and debt are sitting in the business, based on agreed definitions."

Why do buyers like it? Because it makes deals comparable. If you value a business at 10x EBITDA, you're valuing the operating cash flows. You don't want that number to be accidentally inflated because the seller decided to leave a pile of cash in the company. You also don't want to overpay because the seller left behind an unpleasant surprise in the form of debt-like obligations.

Why do sellers accept it? Because it's the market standard in many processes, and because cash and debt are, in theory, measurable. But sellers should treat "measurable" as a dare. Measure it wrong and the bridge becomes a trapdoor.

The best way to read a cash-free, debt-free construct is to ask one question: *What balance sheet items are we economically treating like debt?* The answer varies by deal, and it's one of the most negotiated parts of the economics even when everyone insists they're only arguing about "definitions."

The working capital adjustment: paying for the business you thought you were getting

Buyers don't just buy EBITDA; they buy the ability to operate on day one without starving the business of oxygen. That oxygen is working capital—current assets like accounts receivable and inventory, minus current liabilities like accounts payable and accrued expenses (all defined per the agreement).

The working capital adjustment is a simple idea with a long shadow: the parties agree on a target level of working capital that represents a "normal" amount for the business. If the company delivers more working capital at closing than the target, the seller gets paid more. If it delivers less, the seller gets paid less. In other words, the deal price is adjusted to reflect whether the buyer is receiving a business with normal operating liquidity or a business that has been quietly squeezed.

This is one of the most common sources of post-closing disputes because "working capital" is a concept, not a natural law. Does it include deferred revenue? What about accrued bonuses? Are reserves calculated the same way they were historically? Are receivables written down more aggressively at closing than they were during the year? The agreement should anchor these questions to a specific methodology—often by referencing a sample calculation, a reference balance sheet, and consistent accounting policies. Without that anchor, both sides can be right in their own accounting universe.

The working capital adjustment is also where behavioral incentives get weird. Sellers are motivated to collect receivables aggressively and delay payables to maximize closing working capital—unless the definition punishes that behavior. Buyers are motivated to treat borderline items conservatively and push liabilities into the closing statement—unless the definition prevents it. A well-drafted mechanism doesn't assume good intentions; it assumes rational ones.

The enterprise-to-equity bridge in pictures (where ``headline" becomes ``wire")

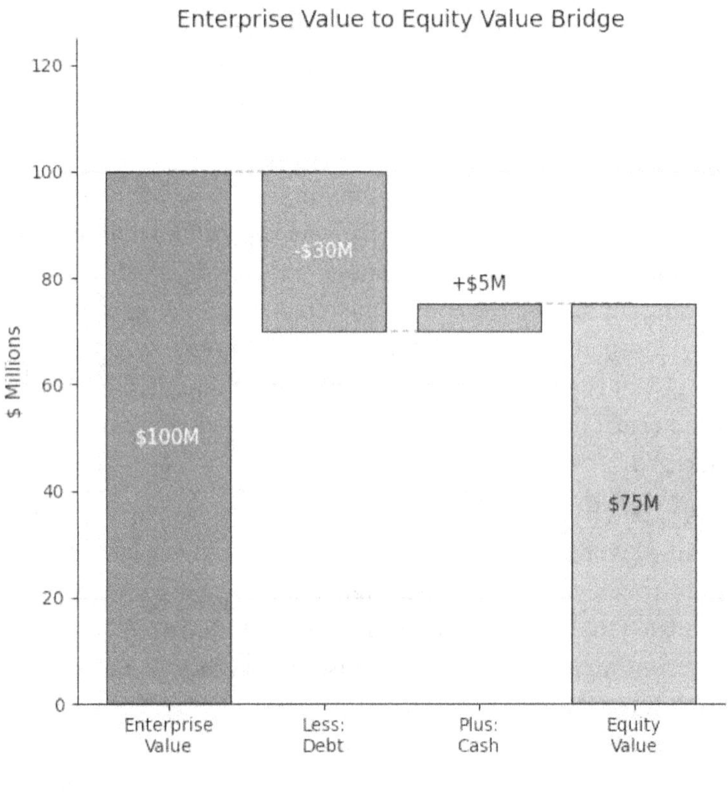

The chart is intentionally simple because the bridge itself is simple; the complexity arrives when humans start negotiating what belongs in each bucket. Two deals can share the same enterprise value and produce wildly different equity checks because one agreement defines "debt" narrowly and the other defines it like a vacuum cleaner.

Why the equity check still isn't ``cash in the seller's pocket"

Even after you bridge from enterprise value to equity value, the seller's day-one cash proceeds can still be lower. Some deductions are obvious: transaction fees, banker success fees, and legal and accounting costs. Some are structural: escrows, holdbacks, and deferred consideration. Some are performance-based: earn-outs.

This is where deal economics becomes less about valuation and more about *risk allocation*. Buyers and sellers often agree on the rough value of the business and then spend most of their time negotiating *when* the seller gets paid and *what* could reduce that payment later.

Escrows and holdbacks: the buyer's collateral for broken promises

An escrow is a portion of the purchase price set aside at closing—often held by a third-party escrow agent—to fund indemnification claims or specific post-closing obligations. A holdback is similar in spirit but sometimes held directly by the buyer rather than a third party. The economic effect is the same: money that would otherwise go to the seller is temporarily withheld.

From a seller's perspective, an escrow can feel like paying for the business twice: "You're buying it, but you're keeping some of my money just in case I lied." From a buyer's perspective, it's the only practical enforcement tool: after closing, the seller may have distributed proceeds, moved on, or become difficult to collect from. An escrow is how the buyer ensures the seller still has skin in the game.

Escrow amounts and durations are the negotiation's way of pricing uncertainty. Cleaner businesses with strong diligence and credible sellers tend to see smaller escrows and shorter periods. Messier situations, or deals where diligence was constrained, tend to see more money held back for longer, or a shift toward insurance products designed to replace large escrows.

Indemnification mechanics: baskets, caps, and survival (a vocabulary of pain management)

Indemnification is the contractual promise that if certain things turn out to be untrue—or certain liabilities arise that the seller agreed to cover—the seller will reimburse the buyer. It sounds straightforward until you realize you need rules for when indemnity starts, how big it can get, and how long it lasts. That's where baskets, caps, and survival periods come in.

A *cap* limits the seller's total liability for certain claims. A *survival period* sets how long a claim can be brought after closing. A *basket* is a threshold: the buyer has to accumulate a certain amount of losses before recovering anything. Basket structures can be deductible-style (buyer absorbs losses up to the threshold) or tipping/first-dollar after the threshold is met (once you cross the line, the seller covers from dollar one). The difference can be economically meaningful and emotionally misunderstood. One feels like a deductible; the other feels like a cliff.

The important point isn't memorizing terms. It's recognizing that indemnification is not an afterthought. It is a slice of the purchase price that remains at risk after closing—sometimes a thin slice, sometimes a thick one—and the economic value of that slice depends on the rules governing it.

Seller notes: when the seller becomes the lender

A seller note is simply a portion of the purchase price paid later as a debt obligation from buyer to seller. It can be used to fill a financing gap, to signal buyer confidence, or to align incentives. It also changes the seller's risk profile dramatically. Instead of taking all cash and walking away, the seller is now exposed to the buyer's creditworthiness and the future health of the business under new ownership.

Seller notes tend to show up when bank financing won't cover the full price, when the buyer is smaller, or when the seller wants to bridge a valuation gap without conceding headline price. They also add a new kind of negotiation: covenants, subordination, security interests, and default remedies. Sellers should treat a note like an investment, not like "almost cash."

Rollover equity: getting paid and staying invested

Rollover equity means the seller reinvests part of their proceeds into the buyer's ownership structure. It's common in sponsor-backed deals where management is expected to remain and grow the business. Economically, it can be attractive: the seller takes some chips off the table while keeping upside if the business scales.

But rollover equity also introduces complexity. The seller is now negotiating not only price, but governance rights, dilution protections, and exit mechanics. The instrument matters: common equity, preferred equity, options, profits interests—each has different economics and different risk. A rollover can be a smart way to align incentives, or it can be a subtle way to shift risk back to the seller under a friendly label.

Earn-outs (contingent consideration): paying for performance that hasn't happened yet

An earn-out is contingent consideration: additional purchase price paid if the business hits specified future targets. Earn-outs exist because humans disagree about the future. The seller believes growth is imminent; the buyer worries the growth is fragile. The earn-out is a compromise: pay for the growth if it actually arrives.

Earn-outs are also relationship stress tests. They require measurement over time and, implicitly, cooperation. Sellers fear the buyer will change strategy, allocate costs differently, or underinvest in the business to avoid paying. Buyers fear the seller will optimize for earn-out metrics at the expense of long-term health, or that the metrics are too easy to manipulate. The more an earn-out depends on accounting judgments (EBITDA add-backs, revenue recognition, allocation of corporate overhead), the more likely it is to become a dispute.

There's a reason accountants treat contingent consideration as technically complex in acquisition accounting: under U.S. GAAP (ASC 805), contingent consideration is generally recorded at fair value at the acquisition date, and then it may be remeasured depending on classification. That accounting reality is a reminder that earn-outs are not a casual side bet—they are a material economic instrument with real-world consequences in reporting, incentives, and litigation risk.

The best earn-outs tend to share a trait: they are built on metrics that match how the business is actually managed and that can be measured without creative writing. If the earn-out requires a committee meeting to decide what "counts," it's already in trouble.

The hidden economics inside ``definitions"

Deal economics are often smuggled into the agreement through definitions that sound neutral.

Net debt might include or exclude lease liabilities, accrued bonuses, unpaid seller expenses, or deferred revenue. *Cash* might exclude restricted cash, trapped cash in foreign subsidiaries, or cash needed to fund specific liabilities. *Transaction expenses* might include only banker and legal fees, or it might include stay bonuses, retention payments, and even certain operating items that are "deal-related" in the buyer's view.

These choices aren't petty. They decide what is paid off at closing, what reduces the equity check, and what becomes a post-close argument. When someone says, "It's just a definition," translate it to: "This is where the money moves."

A short example: the same EV, three very different outcomes

Imagine a business valued at $100 million enterprise value. It has $30 million of debt and $5 million of cash, so the simple bridge suggests $75 million of equity value.

Now layer in common mechanics:

- A $7 million escrow for general indemnity
- $3 million of transaction expenses paid at closing from proceeds
- A $4 million working capital shortfall versus the target
- A $10 million earn-out payable over two years if performance targets are met

On day one, the seller might receive something like:

$$\$75M \text{ equity value} - \$7M \text{ escrow} - \$3M \text{ fees} - \$4M \text{ working capital} = \$61M \text{ cash at closing}$$

And then, maybe, $7 million returns from escrow later (minus claims), and maybe some or all of the $10 million earn-out pays out later (or not). The "$100 million deal" headline is still technically true in casual conversation, but it's a terrible tool for predicting what the seller experiences.

Why LOIs hint at economics early (and why that hint matters)

Even when an LOI is non-binding, it often flags the economic levers: cash-free/debt-free, escrow size, working capital adjustment concept, earn-out framework, rollover percentage. Those are not decorations. They are the blueprint for where negotiation pain will concentrate later. If those items are vague, expect the definitive phase to be expensive in time and trust. If they're specific enough to set expectations, the team can argue about real issues rather than re-litigating the existence of issues.

The LOI is also where the parties decide what kind of risk-sharing relationship they want. A high cash-at-close, low-earn-out offer says, "We'll take the risk and pay you now." A lower cash-at-close with more contingent consideration says, "We'll pay you for what shows up later." Neither is inherently better. The right answer depends on business volatility, the buyer's ability to influence outcomes, and the seller's appetite for continued exposure.

The only question that matters: what is certain, what is delayed, what is conditional?

If you want a deal economics filter that works on any transaction—strategic, sponsor-backed, large, small—use this:

- *Certain*: cash wired at closing that is not meaningfully subject to clawback
- *Delayed*: escrows/holdbacks and deferred payments that are scheduled but still exposed to claims or credit risk
- *Conditional*: earn-outs and other contingent consideration dependent on future performance or events

Every term in the purchase agreement is secretly trying to move dollars from one bucket to another. Buyers push to move dollars from *certain* to *delayed* or *conditional* to protect themselves. Sellers push in the opposite direction to lock in certainty. The negotiation is not abstract. It's a tug-of-war over time and risk, with accounting definitions as the rope.

When a deal feels "expensive" or "cheap," it's often because people are talking past each other—one side is quoting enterprise value, the other is thinking about the equity check, and both are ignoring what portion of that check is actually guaranteed. Get the bridge right, get the definitions right, and you'll be arguing about the business, not about the arithmetic.

1.5 Protecting Information: NDAs and VDRs

In M&A, information is not a courtesy. It's consideration. The seller hands over the private machinery of a business-customer economics, pricing logic, employee compensation, supplier dependencies, maybe even the embarrassing operational duct tape that somehow holds everything together. The buyer uses that information to decide whether to pay, how to structure, and what protections to demand. If that information leaks, the damage is rarely abstract: customers ask uncomfortable questions, competitors sharpen their knives, employees polish resumes, and the seller's negotiating leverage can evaporate in a weekend.

Confidentiality is often described as a legal topic, but in practice it's a security system with two layers. The *NDA* is the contract layer: it defines what the buyer is allowed to do, who is allowed to know, and what happens if the buyer (or someone on the buyer's team) slips. The *VDR* (virtual data room) is the operational layer: it controls access, records behavior, and makes it harder to "accidentally" forward your crown jewels to the wrong inbox. One without the other is either unenforceable or unmanageable.

Why sellers are cautious (and why buyers should respect it)

A seller's fear is not that a buyer will steal a PDF. The fear is that a buyer will steal *optionality*. If customers learn the company might be sold, renewal negotiations change. If employees learn a sale is likely, retention becomes a cost line item. If suppliers learn a new owner is coming, terms shift. If competitors learn which accounts are most profitable, they don't need your entire sales playbook—they just need the page with the margins.

The other fear is more subtle: buyers are often also competitors. Even when they aren't direct competitors, they can be adjacent in ways that matter. A strategic buyer might say they're exploring an acquisition, but the seller can't unknow that the buyer's product team would love to see the seller's roadmap and customer usage data. The NDA and VDR are the seller's way of saying, politely, "Yes, you can look, but you can't *learn* in a way that hurts us if you walk."

Buyers, for their part, are not villains for wanting information. They are underwriting risk, and underwriting requires facts. A disciplined buyer will often volunteer constraints—clean team protocols, limited access lists, segregated competitive personnel—to earn deeper diligence. Sellers tend to interpret that not as weakness but as maturity: a buyer who can control their own organization is more likely to close and less likely to create collateral damage.

The NDA: a short document that carries long consequences

Most NDAs are signed quickly, which is exactly why they deserve more respect than they get. The NDA is the first binding agreement in many deals. It can outlive the transaction, the relationship, and the people who negotiated it. And it often does.

The simplest part of an NDA is also the part people misread: "confidential information." It's rarely limited to documents stamped CONFIDENTIAL. It often includes anything disclosed orally, anything derived from the disclosed information, and sometimes even the fact that discussions are happening at all. That last piece matters. If the NDA treats the existence of negotiations as confidential, a casual comment to a vendor-"We might be acquired"-can be a breach.

A well-drafted NDA is also specific about *purpose*. Purpose language is not there to sound formal; it is there to prevent misuse. If the purpose is "evaluating a possible transaction," the buyer is not supposed to use the information to compete, to recruit employees, or to renegotiate commercial terms with the seller in an unrelated relationship. Purpose is the fence; confidentiality is the lock.

The real pressure points: provisions that change behavior

Some NDA terms are truly routine. Others are where deals get quietly strategic.

Permitted recipients sounds innocuous until you realize it determines whether the buyer can share the seller's information with just an internal deal team, or with a broad group that includes operators, product managers, and sales leaders. Sellers often insist on a narrow list, sometimes excluding anyone involved in competitive decision-making. Buyers sometimes push back because operational diligence requires operational expertise. The compromise is often role-based access through the VDR, sometimes paired with a "clean team" arrangement for the most sensitive information.

Non-solicitation provisions can matter even when the deal is friendly. Sellers don't want buyers using diligence access as a recruiting pipeline. Buyers often want a non-solicit carve-out for general advertising or unsolicited inbound applications. This is one of those clauses that seems small until a key employee quits and everyone tries to remember what they agreed.

Standstill provisions are the NDA's sharpest tool in public-company contexts and sometimes in private deals with strategic buyers. A standstill restricts the buyer from buying shares, launching a hostile bid, or otherwise going around the seller's controlled process. For a seller, it's protection against being "studied" and then attacked. For a buyer, it can feel like being asked to negotiate with one hand tied. The existence of a standstill is a signal: the seller is not just managing confidentiality; they're managing *power*.

Compelled disclosure language addresses subpoenas, court orders, and regulatory requests. Buyers sometimes treat it as hypothetical; sellers shouldn't. This clause defines whether the buyer must give notice, seek protective orders, and cooperate to minimize disclosure. In highly regulated industries, compelled disclosure can be a real path for information to escape.

Return or destruction provisions are another clause people sign and forget. They raise practical questions: What counts as "destroyed" when backups exist? What about emails? What about notes? Many NDAs now acknowledge reality by allowing retention in ordinary-course backups while still requiring confidentiality. The key is not pretending deletion is perfect; it's ensuring retention doesn't become an excuse for later misuse.

The NDA as a filter: seriousness has a signature

Sellers often use the NDA itself as a seriousness test. A buyer who drags their feet on signing an NDA is either disorganized or not truly committed. A buyer who signs quickly but later claims they didn't realize the terms were restrictive signals a different problem: they may not control their internal communication, which is exactly what confidentiality requires.

From the buyer's side, negotiating the NDA can be a way to avoid future friction. If the buyer knows they will need to involve certain operating leaders, it's better to negotiate that access up front than to "accidentally" over-share later. A clean NDA negotiation is often the first proof that both sides can be precise without being hostile-an underrated predictor of how the definitive agreement phase will feel.

The VDR: where confidentiality becomes logistics

If the NDA is a set of rules, the VDR is the building that enforces them. A VDR is not a shared drive with a password. It's a controlled environment designed for high-stakes information sharing: permissions, user groups, watermarking, audit trails, Q&A workflows, and sometimes restrictions on printing, downloading, or screen capture.

The best way to understand a VDR is to see it as a throttle. It allows the seller to increase disclosure as the buyer increases commitment. Early on, buyers might see a few high-level folders: financial summaries, customer concentration, key contracts in redacted form. As the deal narrows—often after an LOI or exclusivity—the seller opens deeper layers: detailed financials, full customer agreements, HR files, IP documentation, compliance materials. This staged disclosure aligns the seller's confidentiality risk with the probability of closing. It's not paranoia; it's rational risk management.

Permissions: the VDR's most powerful economic feature

Permissions are not just about secrecy. They are about speed.

When permissions are well designed, buyers can work in parallel. The tax team sees tax folders. The IT team sees security materials. The HR team sees benefit plans. No one is waiting for someone else to forward documents, and no one is guessing which file is "final-final-2.pdf."

When permissions are sloppy, diligence slows to a crawl. Buyers flood the seller with basic requests because they can't find what they need, or because the wrong people can't access it. Sellers then spend their days playing gatekeeper rather than running the business. That operational distraction is not just annoying; it can degrade performance during the very period the buyer is measuring performance. Poor VDR management can literally reduce purchase price through missed numbers or an earn-out that becomes harder to hit.

Permissions are also political. Sellers often want to restrict competitive personnel-sales leadership, product teams, procurement-to avoid strategic leakage. Buyers argue they need operators to underwrite integration and synergy. A common compromise is "need-to-know" access plus clean team protocols for the most sensitive information (customer-level pricing, margin by account, product roadmap). The VDR makes those compromises executable.

Watermarking and download controls: deterrence beats punishment

A watermark that includes the user's name, company, timestamp, and sometimes IP address looks like a minor annoyance until you imagine a leaked document on someone's desk. Watermarking changes behavior because it makes leaks attributable. Most breaches of confidentiality are not grand conspiracies; they are small acts of carelessness. Deterrence is often more effective than litigation threats, and watermarking is deterrence made visible.

Download controls matter for the same reason. Sellers may allow online viewing but restrict downloads for especially sensitive documents. Buyers sometimes push back because offline work is easier, and because certain diligence providers want to run analyses on files locally. The compromise tends to be selective: allow downloads for financial statements and models, restrict for customer contracts or proprietary technical documents, or allow downloads only for trusted advisors under strict confidentiality obligations.

There is no universal right answer; there is a right answer for the risk profile of the business. A defense contractor, a healthcare provider, and a consumer services business will have very different tolerances for information leaving the room.

Audit trails: the VDR's memory (and occasionally its weapon)

VDR audit logs track who accessed what, when, and for how long. In a cooperative deal, this is mostly useful for project management: if the buyer keeps asking for a document they already viewed, the seller can redirect them. In a less cooperative deal, audit trails can become evidence.

Audit trails matter for two reasons. First, they help sellers measure buyer seriousness. A buyer who claims to be ready to sign but hasn't opened key folders is either reckless or bluffing. Second, they can matter in post-close disputes about disclosure. If a buyer later claims they were never shown a document, the seller may be able to prove it was uploaded and accessed. That doesn't win every argument-lawyers will still debate materiality and interpretation-but it changes the conversation.

The Q&A workflow: where deals either stay professional or get messy

Every deal generates questions. The only choice is whether questions become a controlled workflow or a chaotic flood.

A centralized Q&A function-often built into the VDR or managed through a structured tracker-protects both parties. It prevents duplicated questions. It keeps answers consistent. It creates a record. It allows the seller to route questions to the right internal owner and to respond on a predictable cadence.

The alternative is email sprawl, where different buyer team members ask different seller team members the same question in slightly different ways, and the seller accidentally gives slightly different answers. Those inconsistencies aren't always lies; they're often just human variation. But in M&A, inconsistency becomes a negotiation opportunity. A buyer who senses confusion starts drafting broader protections. A seller who senses a fishing expedition starts withholding. The Q&A workflow is a cheap way to preserve trust.

One practical best practice: require questions to be specific and to cite the document or folder prompting the question. Vague questions invite vague answers, and vague answers invite more questions. Precision compounds.

Redaction and clean rooms: the art of showing enough

Some information is too sensitive to share broadly even under NDA: customer-level pricing, patient data, source code, trade secrets, certain security documentation. Sellers handle this in a few ways.

Redaction is the simplest: remove customer names or pricing terms while preserving structure. It's useful early but has diminishing returns; buyers eventually need the real data to underwrite.

Clean rooms (sometimes physical, often controlled virtual access) are the more serious tool. A clean room might allow viewing but no copying, or access only to a small "clean team" of people who are walled off from competitive decision-making. Clean rooms are often used when the buyer is a competitor or when regulatory or privacy constraints are tight.

The key is alignment with deal probability. Sellers shouldn't hand over the nuclear codes to a buyer who hasn't even committed to a price. Buyers shouldn't demand nuclear codes if they aren't willing to take the obligations seriously. This is where deal teams earn their keep: designing an information ramp that supports diligence without creating avoidable risk.

Common failure modes (and how they quietly kill deals)

Confidentiality failures don't always look like a scandal. They often look like a vibe change.

A leak to employees can create a sudden wave of "just checking in" recruiter messages, which then becomes retention payments and distraction. A leak to customers can trigger rebids or delayed renewals. A leak to competitors can change market behavior. Even if the deal closes, the buyer ends up acquiring a weakened asset-the corporate equivalent of buying a car after someone siphoned the gas.

Operational failures are just as damaging. A VDR that is incomplete or disorganized creates buyer skepticism. Skepticism creates slower decision-making. Slow decision-making creates more time for the business to wobble. And wobble is expensive because it shows up in the exact metrics buyers underwrite: revenue growth, churn, margins, and working capital.

The other failure mode is over-protection. Sellers can lock down the VDR so tightly that serious buyers can't do their work, which creates a self-fulfilling prophecy: buyers reduce their price to compensate for uncertainty or walk away due to perceived opacity. The best confidentiality strategy is not maximum secrecy; it's *calibrated disclosure.*

A dealmaker's rule of thumb: match information depth to commitment

The most successful processes treat information like a staircase, not an open door. Early materials help buyers decide whether to lean in. Deeper materials are earned by credible bids and increasing commitment. The deepest materials-customer-level economics, sensitive employee data, security details-are shared only when the buyer is close enough to closing that walking away would be costly.

This approach isn't manipulative; it's efficient. It protects the seller from unnecessary exposure and protects the buyer from spending weeks drowning in documents before knowing whether they have a real shot at winning.

In the end, NDAs and VDRs are less about distrust and more about professionalism. They make it possible for strangers to exchange the most sensitive truths about a business without destroying it in the process. When confidentiality is handled well, the deal team spends its time arguing about value and risk in the open. When it's handled poorly, the deal team spends its time managing fallout—and fallout has a way of getting priced into the deal, one painful concession at a time.

Chapter 2

Crafting the Letter of Intent (LOI)

Think of the Letter of Intent (LOI) as the engagement ring of the M&A world-it signals serious intent and takes the seller off the market, but it isn't the wedding vows just yet. This chapter walks you through drafting this critical document, ensuring you don't accidentally legally bind yourself to a price before you've seen the skeletons in the closet, while securing the exclusivity you need to spend money on due diligence.

2.1 Process Dynamics: Auctions vs. Bilateral Deals

Walk into an auction and you can feel it in the calendar invites. Back-to-back management calls. A data room that looks like someone dumped an accounting department into a filing cabinet and hit "share". A "final bid" deadline that arrives suspiciously fast for something labeled *final*. The seller is not just selling a company; they're selling *tempo*. And in M&A, tempo is leverage.

A bilateral deal is the opposite kind of power. It rarely announces itself with a process letter and a bid date. It starts as a conversation that becomes a meeting that becomes a "let's see if there's something here". The leverage isn't in speed; it's in *optionality*—the buyer's ability to slow down, ask for one more dataset, meet the second-layer managers, and keep the whole thing in a private lane. Where auctions weaponize time, bilateral negotiations weaponize *patience*.

The LOI sits right in the blast radius of this difference. Same document title, completely different physics. In an auction, an LOI is often a *ticket* you buy to stay in the stadium: you write something crisp enough to pass the seller's filter, and you accept that many of the details will be litigated later in definitive documents. In a bilateral deal, an LOI is more like a *trapdoor*: once you step through, you've usually agreed to rules (especially exclusivity) that change what you can do next. One environment rewards speed and signaling. The other rewards control and protection.

Who owns the process owns the leverage

A seller running an auction is trying to do one simple thing with very complicated tools: make each bidder fear losing. Not fear losing the company in the abstract—fear losing it to *someone else*. That fear changes behavior. It makes buyers tolerate seller-friendly terms, accept short diligence windows, and write LOIs that feel more like promissory notes than proposals.

The mechanism is straightforward. The seller (usually through an investment bank) standardizes the steps: teaser, confidentiality agreement, information memorandum, first-round indication of interest, management presentation, data room access, second-round bid, maybe a confirmatory diligence sprint, then "best and final". Standardization is not just for cleanliness. Standardization makes bidders comparable—and comparability makes it easy to say, "You're behind."

In a bilateral deal, there's no scoreboard. There might not even be a shared definition of "ahead". That ambiguity gives each party room to maneuver. A buyer can explore structure alternatives without looking indecisive. A seller can float valuation expectations without having to defend them against three other offers. The leverage is softer, but it's real: the party who can credibly *wait* usually gets to decide what "reasonable" means.

The two kinds of competition that matter

When people say "auction", they usually mean "multiple bidders". But the buyer isn't just competing against other buyers. The buyer is also competing against the seller's *fallback plan*. That fallback plan might be "run this business for another year and try again", "do a dividend recap", "bring in a minority investor", "merge with a strategic partner", or "go public". You don't need to know the exact plan to feel its presence. If the seller has alternatives, they can insist the process stays strict. If they don't, the auction starts to look like theater.

A buyer should try to diagnose which competition is real. Real bidder competition shows up in the form of clean deadlines, controlled access, and limited exceptions: you can't talk to the CFO twice, you can't request custom schedules, you can't ask for a bespoke diligence timeline. Real fallback-plan competition shows up differently: the seller is calm about timing, less reactive to "we need more information", and more willing to let a bidder drift without chasing them. That calmness is often the tell that the company is not *for sale at any price*; it's for sale at a price and set of terms that beat the alternative.

In a bilateral deal, the fallback plan is always present because there are no other bidders to keep everyone honest. The seller can "shop" quietly, even if they swear they're not. The buyer can do the same. The negotiation is less about outrunning rivals and more about preventing the other side from quietly creating them.

Auctions: why speed feels like certainty (even when it isn't)

An auction compresses time, and compressed time creates a very specific psychological trick: it makes weak information feel actionable. You get 48 hours to respond to a detailed Q&A log. You receive a customer list but not the full contracts. You see monthly financials but not the revenue recognition memos. You're asked to "get comfortable" with working capital without seeing how the company defines it. In normal life, you would never make a high-stakes decision on that basis. In an auction, people do it all the time, because the alternative is getting cut.

Sellers understand this. They don't need to hide the ball; they just need to decide *when* you're allowed to see it. Controlled disclosure is a form of leverage. If you only learn the messy bits after your LOI, your LOI becomes a commitment device. You now have internal momentum: you've told your IC (investment committee) a story, you've aligned your deal team, you've started thinking about integration, you've implicitly signaled seriousness to lenders. The seller's goal is not to trick you; it's to make you invest enough that walking away is painful.

That pain changes the LOI you write. Buyers in auctions often overemphasize headline price and underemphasize mechanics—the small print that decides who pays for surprises. It's not irrational; it's triage. If the seller is selecting a finalist primarily on price and perceived certainty, buyers focus on what is scored.

A savvy buyer still uses the LOI to stake out the mechanics that matter most, but does it with a particular style: short, crisp, non-negotiable where it must be, and deliberately "market" where it can be. Auctions punish long legal essays. They reward clean terms that look like they'll draft quickly.

Bilateral deals: the quiet leverage of being the only dance partner

In a bilateral negotiation, the buyer's best weapon isn't aggression. It's the ability to say, politely and credibly, "We're not going to guess." You can insist on diligence sequencing: "We'll move to an LOI after we see customer concentration by cohort", or "We'll need a quality of earnings scope agreed before we lock a purchase price mechanism." You can ask for a management dinner without it being part of a choreographed roadshow. You can take time to understand the business like an owner would, not like a bidder would.

The seller's leverage is different. It's not "someone else will take your place" (though they can always try to create that). It's informational: the seller knows the business in their bones, knows which issues are explainable, which are embarrassing, and which are landmines. In bilateral deals, sellers sometimes attempt to monetize this by pushing for early exclusivity: "We'll open the books, but only if you stop talking to anyone else." That request can sound reasonable. It's also a pivot point where leverage shifts dramatically.

If the seller gets exclusivity early, the buyer loses the pressure of competition but gains a different pressure: the pressure of *time invested*. Once you are exclusive, you are building the seller's deal, whether you close or not. Every week you spend in diligence is a week the seller isn't taking other calls. That creates an emotional dynamic that can turn a rational negotiation into a moral argument: "You've had us exclusive for 60 days; you can't change the price now." A buyer in a bilateral deal should expect that line, and should write the LOI in a way that makes the rules of the game explicit *before* emotions start drafting the contract.

The seller's playbook in an auction (and how it shows up in your LOI)

Sellers in auctions typically optimize for three things: price, certainty, and speed. The LOI is where bidders are forced to translate promises into a format that can be compared side-by-side. That comparability is why auction LOIs often start to look similar: same headings, same sequence, same "clean" structure. When everyone's LOI is formatted like a standardized test, the seller gets to grade quickly.

The first seller tactic is to create a "default" markup. Sometimes it's explicit: the seller circulates an LOI template or a draft purchase agreement and asks bidders to mark it. Sometimes it's implicit: bidders are told to "limit exceptions" and "keep comments high-level." Either way, the seller is trying to anchor the negotiation around seller-preferred terms by making deviation costly.

The second tactic is to measure seriousness through *friction*. They might ask for a fully financed bid, a committed equity check, evidence of board approval, or a near-final diligence plan. The goal is to separate bidders who *can* buy from bidders who are still trying to decide if they want to.

The third tactic is to keep optionality until the last responsible moment. A seller wants multiple bidders advancing as long as possible because each bidder is a lever on the others. That means the seller will often delay granting exclusivity until a "winner" is chosen, and sometimes even after, by labeling it "confirmatory diligence" and keeping the runner-up warm.

All three tactics affect how a buyer should craft an LOI in an auction. Your LOI isn't just a summary of economics; it's a competitive artifact. It should read like something that can close, not something that needs a committee to interpret it.

The buyer's playbook in an auction: win without overpaying in the small print

Buying in an auction is like bidding on a house where you're allowed to tour it, but only for ten minutes, and the seller tells you the inspection can happen after you put down a nonrefundable deposit. The rational response is not to refuse to bid. The rational response is to decide which risks you can price, which risks you must eliminate, and which risks you can't tolerate at all.

Buyers often talk about "winning the auction" as though it's purely a price game. That's a mistake. Auctions are frequently won on a blend of economics and certainty. Certainty is not just "cash versus financing." It's also: Are you asking for a financing out? Are you asking for a broad diligence out? Are you proposing a structure that requires third-party consents that will be slow or uncertain? Are you pushing a complicated earnout? Are you insisting on indemnity terms that will take weeks to negotiate?

A practical way to think about the auction LOI is to treat it like a menu where you have a limited number of "customizations" before the kitchen stops taking your order. Spend those customizations on terms that move real dollars or prevent real disasters. Examples include: the purchase price mechanism (especially cash-free/debt-free and working capital concepts), treatment of transaction expenses, and the boundaries of what diligence can do to price. If you waste your customization budget on cosmetic language, you will still lose the terms that actually matter.

That doesn't mean being vague. In auctions, vagueness is a tax that shows up later as either (a) re-trading accusations or (b) seller-friendly defaults. The art is to be precise where the stakes are high and flexible where the seller is grading for speed.

Bilateral negotiations: leverage is usually earned, not seized

In a bilateral deal, the most common mistake is confusing friendliness with alignment. The tone may be warm. The meetings may be collaborative. There may be genuine admiration on both sides. None of that changes the economic truth that buyer and seller are negotiating against each other. The difference is that the negotiation doesn't have an audience. Without other bidders, each side can afford to be patient and polite—and that politeness can lull teams into leaving key leverage points unaddressed until they become crises.

Bilateral leverage is earned through preparation and clarity. A buyer earns leverage by showing they understand the business well enough to move efficiently: crisp diligence requests, fast turnaround on questions, decision-makers present. Sellers grant more flexibility to buyers who look like they'll close. Conversely, a buyer who drifts invites the seller to manufacture competition—sometimes by actually calling other buyers, and sometimes by simply implying that they might.

Sellers earn leverage in bilateral deals by controlling access. They can decide which customers you can reference, which managers you can meet, which projections you can rely on, and how quickly the data room fills up. They can also choose the narrative: "We're growing; we're not desperate; we're choosing a partner." A buyer needs to respect that narrative while still insisting on verification. The LOI becomes the place where both sides decide whether verification is a right or a favor.

Exclusivity is the fulcrum: why it means different things in auctions and bilateral deals

Exclusivity (often called a "no-shop") is where leverage shows up as a contract clause instead of a vibe. In an auction, exclusivity usually arrives late, after the seller has extracted competitive tension. It tends to be shorter, because the seller has already run much of the process and doesn't want to be handcuffed for long. The buyer still wants it, because the buyer is about to spend serious money on confirmatory diligence, legal drafting, and lender work. But the buyer's leverage to demand a long exclusivity window is weaker because the seller can credibly say, "You're not the only option."

In a bilateral deal, exclusivity is often requested early. That early request is a rational seller move: if there's no auction, the seller wants to prevent the buyer from using the seller's openness as a free option while simultaneously shopping. From the buyer's perspective, early exclusivity can be attractive because it turns a fuzzy negotiation into a focused one. The danger is that the buyer signs away optionality before the buyer has enough information to price the business confidently.

The key insight is that exclusivity does two different jobs. In an auction, exclusivity is primarily about protecting the buyer's *closing work* (documentation, financing, confirmatory checks) from being wasted by a seller who keeps shopping. In a bilateral deal, exclusivity is also about protecting the seller from being used as a *price discovery service*. Those are both reasonable goals. The fight is about how much time and flexibility each side gives up to achieve them.

The hidden leverage in deadlines: ``drop-dead dates'' and manufactured urgency

Deadlines in M&A have a funny habit of being both arbitrary and decisive. Auction deadlines are often real in the sense that the seller will, in fact, stop taking calls after the date. They're arbitrary in the sense that the seller chose the date to create urgency, not because the business will collapse if bids arrive on Tuesday instead of Monday.

Buyers should treat deadlines as signals. A deadline signals the seller's confidence in the process, the level of buyer interest, and the seller's willingness to trade off price for speed. When a seller is truly oversubscribed with interest, deadlines are rigid and consequences are immediate. When interest is thin, deadlines start to "move", and exceptions quietly appear. A buyer can't always see the book, but they can read the clock.

In bilateral deals, deadlines are often invented for internal alignment: "We need an LOI by the end of the month because our board meeting is then", or "We need to sign before quarter-end". Sometimes those are real constraints. Sometimes they are negotiation tools designed to force the other side to commit before they're ready. The buyer's leverage is the ability to test whether the deadline is real by proposing an alternative: "We can meet that date if we limit diligence to X and reserve Y as a post-LOI confirmatory item." If the seller accepts, the deadline was negotiable. If the seller refuses, the deadline might be real—or the seller might be using it to improve terms. Either way, the LOI becomes the written record of what you agreed to do under time pressure, which is exactly why it must be drafted like it will be read later by someone unfriendly.

Information asymmetry: auctions ration it; bilateral deals drip it

In auctions, information is rationed by design. Sellers may release certain schedules only to second-round bidders. They may delay full customer contracts until a bidder is near exclusivity. They may provide a management presentation that is polished but not particularly interrogable. The buyer's team ends up building a deal model on partial information, with ranges and sensitivities standing in for facts.

In bilateral deals, information is often delivered as a drip. The seller might respond to requests in batches, or add documents as questions arise. This feels more collaborative, and sometimes it is. But the drip can create its own leverage problem: it makes it harder for the buyer to declare diligence "complete". If information arrives continuously, the seller can argue that the buyer's requests are endless, or the buyer can feel justified in extending the timeline indefinitely. Both outcomes breed distrust.

A strong LOI strategy acknowledges this dynamic. Auction LOIs should specify what needs to be true for the buyer to move quickly (access, responsiveness, and a clear path to confirmatory diligence). Bilateral LOIs should define a diligence plan that is ambitious but bounded, with clear expectations around turnaround times and access to key people. Not because you want to be rigid, but because you want to prevent the deal from becoming a slow-motion argument about what "reasonable" means.

``Certainty" is not a personality trait; it's a set of choices

Sellers love "certainty". Buyers love "flexibility". Those preferences show up as moral language, which is how bad negotiations happen. The seller frames certainty as seriousness and flexibility as flakiness. The buyer frames flexibility as prudence and certainty as recklessness. Both framings are self-serving. The truth is simpler: certainty is a set of choices about where risk sits between signing and closing.

In an auction, buyers often signal certainty by removing conditions, offering a cleaner structure, minimizing contingencies, and proposing a fast signing timeline. Strategic buyers sometimes have an edge here because they can pay cash and don't need lender approvals, but they also have their own uncertainty: regulatory approvals, integration risk, internal governance, and sometimes antitrust. Financial sponsors can be highly certain if they have committed financing or a strong relationship with lenders, but sellers will scrutinize the "outs" in the financing package even if the LOI looks clean.

In bilateral deals, certainty is negotiated more explicitly. A buyer can say, "We will be certain on price if you are certain on disclosures", or "We will move fast if we have weekly access to the CFO and controller." That kind of trade can be hard to execute in an auction because the seller won't give bespoke accommodations to one bidder. In bilateral negotiations, it's often the heart of the bargain.

Re-trading: the sin that auctions punish and bilateral deals invite

Re-trading is what sellers call it when the buyer tries to change economic terms after the LOI, usually because diligence revealed something unpleasant or because the buyer's enthusiasm cooled. Buyers call the same thing "updating the deal to reflect reality". The difference is not semantic; it's leverage.

Auctions punish re-trading because the seller can credibly say, "We'll go back to the runner-up." That threat is not always real, but it is often real enough. Buyers know this, which is why auction LOIs tend to be more aggressive and why buyers sometimes take more risk than they would prefer. Re-trading in an auction can also damage reputation with bankers, who have long memories and short patience.

Bilateral deals invite re-trading because there is no visible runner-up. Once exclusivity is signed, the buyer may feel they have leverage to push down price or tighten terms, especially if diligence uncovers issues that were not disclosed. The seller, having stopped talking to others, is now exposed. They may have told employees, or at least the leadership team. They may have delayed strategic initiatives. They may have mentally spent the proceeds. That exposure makes the seller vulnerable—and that vulnerability is why sellers fight hard over exclusivity duration and over what counts as a legitimate basis to change terms.

A disciplined buyer doesn't rely on re-trading as a strategy. The cleanest wins come from writing an LOI that anticipates how diligence will change certainty and builds an agreed mechanism for handling surprises: either through purchase price adjustments, escrow/holdback concepts, specific indemnities later, or clear conditions that are understood to be real. The point is not to lock everything down early; it's to prevent the inevitable surprises from turning into a character trial.

How to read leverage in the seller's behavior

Leverage leaks out in small decisions long before it shows up in big clauses.

If the seller insists on a single, narrow window for management Q&A, they have leverage (or are trying to project it). If they offer flexible scheduling and extra calls, they may be accommodating—or they may be nurturing interest.

If the seller provides clean, organized data room materials early, they are confident and prepared. If the data room is messy and late, it may signal either disorganization or a deliberate attempt to keep bidders from digging too deep before they commit.

If the seller's adviser is willing to discuss process exceptions ("we can give you one extra week"), that usually means the seller values your participation. In an auction, being valued is a form of leverage—for you. It doesn't mean you can rewrite the rules, but it does mean you can choose which rule fights are worth having.

In bilateral negotiations, similar signals exist but are easier to misread. Responsiveness, openness, and friendliness are not the same as weakness. Sometimes they indicate a seller who is confident enough not to posture. The only reliable way to interpret them is to watch what happens when you ask for something that costs the seller optionality: a longer diligence period, broader access, or a delay in exclusivity. The reaction to those requests tells you who is actually holding the pen.

What ``good" looks like: matching LOI posture to process reality

The best LOI posture is not universally buyer-friendly or seller-friendly. It is *process-aware.*

In an auction, a buyer should write an LOI that is fast to evaluate and hard to misunderstand. Lead with the headline economics, but do not neglect the mechanical terms that determine whether the headline is real. Use plain language where possible. Avoid exotic structures unless they create value that the seller will actually credit. Above all, signal that you can close: show that decision-makers are aligned, financing is credible, diligence is organized, and the path to a definitive agreement is short.

In a bilateral deal, a buyer should use the LOI to create a protected workspace: enough exclusivity to justify serious effort, enough diligence access to avoid guessing, and enough clarity on key mechanics to prevent the definitive agreement phase from becoming a second negotiation about first principles. The buyer can afford to be more precise because speed is not being scored the same way. The seller can afford to negotiate because there is no deadline designed to compare five offers on a single page.

The seller's matching posture has its own logic. In auctions, sellers should guard against process fatigue and confidentiality leaks by keeping the field tight and the steps clear. In bilateral deals, sellers should guard against being "stalled" by demanding a real timetable and meaningful milestones, not just vague promises.

Leverage, in other words, is not something you declare. It's something the process hands you, and then takes away, depending on how you behave. The LOI is where that transfer happens in writing.

2.2 Anatomy of an LOI

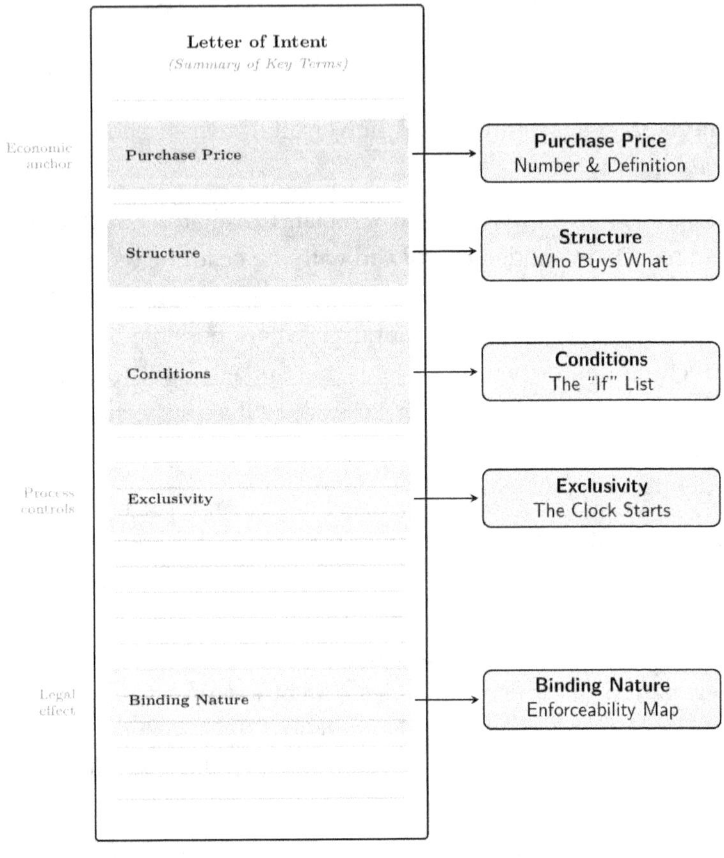

An LOI looks deceptively polite. A few pages. A calm tone. Phrases like "contemplates" and "subject to." Then you watch what happens after it's signed: bankers start booking daily calls, lawyers start billing in meaningful units of time, lenders want diligence readouts, and operating executives start acting like closing is inevitable. That's why the real function of an LOI isn't to *finalize* a deal. It's to *choose*

the arguments you're going to have later—and to decide which of those arguments are already over.

The fastest way to read an LOI is to ignore the paragraphs and hunt for five pressure points: price, structure, conditions, exclusivity, and binding nature. Those aren't just headings; they're levers. Price decides what the buyer is willing to pay in a perfect world. Structure decides what the buyer is actually buying (and what liabilities hitch a ride). Conditions decide how easily either side can walk away with a straight face. Exclusivity decides who gets to negotiate in private. Binding nature decides what happens when someone claims, afterward, that they "didn't mean it like that."

The LOI's split personality: economics vs. process

Most LOIs are hybrids: half business memo, half process contract. The business memo portion sketches the economic bargain—what you'd tell your investment committee if you had to explain the deal on one slide. The process contract portion governs behavior while the parties figure out whether that bargain survives contact with reality.

This split is why LOIs create confusion inside companies. A CEO reads "non-binding" and thinks it's safe. A deal team reads the no-shop and realizes they've just agreed to run a marathon with their shoelaces tied together. Both reactions are rational. The trick is to treat the LOI like a document with two different *audiences*: (1) the future drafting teams, who need a clean blueprint for definitive agreements, and (2) the future litigators (or at least the future angry counterparty), who will weaponize whatever is written when something goes sideways.

Purchase price: the number that matters less than the definition

The headline price is the most quoted term in an LOI and, routinely, the least understood. "$200 million" sounds precise. It rarely is. The real price is not a number; it's a *recipe*. What goes into the recipe determines whether the buyer pays $200 million or $200 million plus a pile of debt, unpaid bonuses, stale inventory, and last quarter's sales tax problem.

At LOI stage, sophisticated parties try to lock down the price *framework* rather than every last dollar. The framework answers questions like: Is the purchase price stated on a cash-free, debt-free basis? Is there a working capital adjustment (a post-close true-up based on the level of working capital delivered at close)? Are transaction expenses paid by the company before close (reducing value to the buyer), or by sellers out of proceeds? Is there an earnout (a contingent payment tied to future performance)? If you only read the first line that says "Purchase Price," you're reading the headline and skipping the footnotes that move the stock.

This is where auctions and bilateral deals feel different in practice. In an auction, buyers often keep price mechanics "market" in the LOI to avoid looking difficult. In a bilateral deal, a buyer has more room to say, early, "We can agree on a price, but we need a shared definition of what we're paying for." That definition is not pedantry; it's a way of preventing the deal from being renegotiated later under the emotionally charged banner of "re-trading."

A useful mental model: the LOI price section is less like a check and more like a label on a bottle. If the label doesn't list the ingredients, you shouldn't be surprised when you taste something you didn't expect.

Structure: the deal's skeleton (and the source of many late surprises)

Structure answers a blunt question: what is the buyer acquiring, exactly? The common options are stock (buy the shares or equity interests), asset (buy selected assets and assume selected liabilities), or a merger (often a legal wrapper around a stock-style acquisition). At LOI stage, structure is often described in a sentence or two, which is just enough to be dangerous.

Why does structure matter so much? Because structure decides which liabilities the buyer inherits by default, how contracts move (or don't), how employees transfer, and how taxes land. Two deals can have identical headline prices and radically different net outcomes because one is structured to leave legacy liabilities behind while the other drags them into the buyer's house like muddy boots.

For a buyer, the structure paragraph is also a credibility test. If the LOI proposes an asset deal when the target has hundreds of customer contracts with anti-assignment clauses (clauses requiring consent to transfer), the seller will read it as either naiveté or gamesmanship. If the LOI proposes a stock deal but the business has known legacy exposures that the seller expected to carve out, the buyer may be walking into a fight that could have been avoided with a cleaner early statement: "Stock purchase, subject to customary pre-closing reorganizations and carve-outs."

In bilateral deals, structure can be negotiated like adults negotiating a real-world constraint: "We prefer an asset deal for liability reasons, but we know your customers won't consent quickly, so here's an alternative." In auctions, structure is often a signaling device: "We will do what is simplest for the seller," which usually means a stock deal for private companies and a merger structure for public ones. The price of that simplicity is that the buyer must later manage risk through representations, warranties, indemnities, escrows, and covenants. Structure doesn't eliminate risk; it relocates it.

Price and structure are not separate clauses; they are a single bargain

The cleanest LOIs make an implicit truth explicit: price and structure are intertwined. A buyer may pay more for a structure that reduces uncertainty. A seller may accept less in exchange for a structure that avoids taxable gain at closing or that keeps certain liabilities with the company rather than the sellers.

This interplay is why LOIs sometimes trigger internal confusion: finance teams think in price, legal teams think in structure, and tax teams think in after-tax proceeds. The LOI has to be legible to all three mindsets or it becomes a battlefield where each function insists the other "didn't understand the deal."

If you want a practical test of whether an LOI is coherent, ask a simple question: could two smart people read this LOI and still disagree about whether the buyer is paying off the company's debt at closing? If the answer is yes, the LOI is not a map; it's a vibe.

Conditions: the "if" that decides whether the LOI is a commitment or a placeholder

Conditions in an LOI are where seriousness gets negotiated. Every buyer wants an out. Every seller wants to believe there isn't one. The LOI is where those competing fantasies meet paper.

Conditions usually include some combination of: satisfactory completion of due diligence, negotiation and execution of definitive agreements, financing (if the buyer needs debt), required consents (customers, landlords, counterparties), regulatory approvals, and internal approvals (board, investment committee, sometimes shareholders). Most of these are reasonable. The tension is not whether they exist; it's how they are drafted.

A broad diligence condition ("satisfactory due diligence in buyer's sole discretion") is comforting to a buyer and terrifying to a seller because it reads like a free option. A narrowly framed diligence condition ("confirmatory diligence consistent with information provided") sounds better to a seller but can be dangerous to a buyer if the information provided is incomplete or misleading. The best drafting often lives in the awkward middle: diligence is a condition, but it's paired with an access commitment, a timeline, and an understanding of what categories are critical (customer concentration, quality of earnings, cybersecurity, environmental, whatever is real for the business). That transforms diligence from a vague escape hatch into a defined process.

In an auction, conditions are scored like blemishes. Every extra condition makes you look less certain. Buyers respond by stripping conditions down to the ones they cannot live without. In a bilateral deal, conditions can be a way to create shared expectations: "We'll move fast, but we need these gates." Either way, conditions tell you what the buyer is afraid of and what the seller is willing to tolerate.

Exclusivity: a single paragraph that can cost millions

Exclusivity is often a short clause and a long shadow. Once signed, it changes the negotiation from a market process to a private conversation. That shift is why exclusivity can be the most valuable thing exchanged in an LOI even though it doesn't look like money.

From the buyer's perspective, exclusivity is protection for spending: third-party diligence fees, legal fees, internal time, lender engagement, integration planning. Nobody wants to fund a seller's auction preparation while someone else wins the deal. From the seller's perspective, exclusivity is risk: the seller pauses other conversations and ties their fate to a single buyer who might discover reasons to renegotiate.

This is where the LOI stops being a summary and starts being a weapon. The duration of exclusivity, its scope (no solicitation, no discussions, no furnishing information), and its carve-outs determine how tight the handcuffs are. In a competitive process, the seller will push for short and narrow. In a bilateral process, a buyer may push for longer because they know there is no built-in competitive urgency forcing the seller to stay engaged.

Even when the LOI is mostly non-binding, exclusivity often is not. That asymmetry is intentional. It's the buyer's insurance policy against being used as a stalking horse, and the seller's lever to keep the buyer moving.

Binding nature: the clause that decides what ``non-binding" really means

Most LOIs contain a paragraph that says, in some form, "This LOI is non-binding except for Sections X, Y, and Z." That paragraph is not boilerplate. It is the boundary wall between "we were negotiating" and "we had a contract."

The binding list typically includes confidentiality (sometimes by reference to an NDA), exclusivity, expenses (who pays what), governing law/venue, and sometimes a commitment to negotiate in good faith or proceed diligently. The non-binding bucket typically includes price and structure. But reality is messier than the buckets. A poorly drafted LOI can accidentally create binding obligations where the parties didn't expect them, or can create just enough ambiguity to invite a fight about what was promised.

If you want to see the stakes, imagine a buyer who signs an LOI, gets exclusivity, then slows down. The seller claims the buyer is dragging their feet in bad faith. The buyer claims they're doing diligent diligence. Whether that becomes a commercial disagreement or a legal one depends heavily on what the LOI says about timing, obligations, and what "good faith" means in context. The binding nature clause is not about legal theory; it's about keeping business arguments from turning into courtroom arguments.

A practical way to scan an LOI in under five minutes

Deal teams often read LOIs like novels, line by line, absorbing tone. That's a fine way to appreciate the prose and a terrible way to spot risk. A better approach is to scan for the parts that control money, then the parts that control time.

Money-control terms include: purchase price definition (cash-free/debt-free, working capital), treatment of debt, treatment of transaction expenses, earnouts or seller notes, rollover equity (if sellers reinvest), escrow/holdback concepts (even if only high-level), and any mention of tax treatment that affects net proceeds. These are the terms that decide whether the deal you model is the deal you pay.

Time-control terms include: exclusivity duration, diligence timeline, access commitments (who you get to meet and what you get to see), signing target, and any drop-dead dates. These are the terms that decide whether the deal stays alive long enough to close.

And then there's the "fight insurance" layer: binding nature, governing law, confidentiality/publicity, and expenses. Those don't make the deal better, but they determine how expensive it is when the deal gets worse.

Making the LOI do real work: clarity that saves legal fees (and emotional bandwidth)

A vague LOI doesn't avoid negotiation; it postpones negotiation into a more expensive phase. Once lawyers are drafting definitive agreements, every ambiguity becomes a markup. Every markup becomes a call. Every call becomes a scheduling problem between busy people who are now emotionally invested. The billable hour doesn't just bill money; it bills attention.

The LOI can prevent that by being crisp about a few core mechanics without trying to draft the entire acquisition agreement in miniature. There's a sweet spot where the LOI is detailed enough to prevent surprises and simple enough to survive the reality that diligence will change someone's mind about something. Hitting that sweet spot is not about legal artistry. It's about choosing plain language for the parts that matter and resisting the temptation to turn every possibility into a paragraph.

One surprisingly effective technique is to write LOI economics as definitions rather than aspirations. "Debt" means *all* indebtedness for borrowed money plus capital leases plus unpaid interest plus change-of-control bonuses, or it doesn't. "Working capital" includes deferred revenue, or it doesn't. "Transaction expenses" include the banker fee, or they don't. If the LOI forces the parties to agree on definitions early, the definitive documents become implementation rather than reinvention.

A compact reference: what each core component is really doing

LOI Component	*What it really controls*
Purchase Price	The economic headline *and* the math underneath it (what cash changes hands at close and why).
Structure	What is being acquired, what liabilities come along by default, and how hard it will be to move contracts and employees.
Conditions	How either party can say "no" later without admitting they changed their mind; also the timeline's choke points.
Exclusivity	Whether the negotiation happens in public (market) or in private (one-on-one), and who has the right to create alternatives.
Binding Nature	Which promises survive "this is non-binding," and therefore which disputes become enforceable obligations.

An LOI doesn't need to be long to be powerful. It needs to be legible in the ways that matter: legible to finance, legible to legal, legible to lenders, and legible to the version of each party that will exist three months later, stressed and tired, trying to remember what was "basically agreed." The anatomy is simple. The consequences of getting it wrong are not.

2.3 Binding Terms and Exclusivity

The LOI's most dangerous sentence is usually a comfort blanket: "Except for Sections __, __, and __, this Letter of Intent is non-binding." People read that and exhale. Then they sign it. Then they behave as if they're engaged. Then someone tries to date other people.

That's the core paradox: an LOI can be mostly non-binding and still be *highly enforceable where it counts*. The enforceable parts aren't the romantic ones (price, synergies, vision). They're the housekeeping rules that govern what happens while everyone spends money trying to get to a definitive agreement. Think of them as the deal's *rules of the hallway*: who can talk to whom, who can say what, who pays for the mess, and whether either side is allowed to quietly look for a better option.

The headline economics in an LOI are often written like a handshake across a table. The binding provisions are written like a fence with a padlock. Confuse the two and you get the worst kind of surprise: not a valuation surprise, but a *behavior* surprise. The seller thinks "we're exclusive." The buyer thinks "we're still exploring." The lawyers wake up.

``Non-binding" is not a force field

The phrase "non-binding" doesn't sterilize the document. Courts don't ignore ink just because the parties used a comforting label. If an LOI contains clear promises that look like a contract—especially promises about exclusivity, confidentiality, or negotiating conduct—those promises can be enforceable even if the rest of the LOI reads like a draft of a draft.

This is not an academic distinction. It changes how you should write, and how you should behave. If the LOI says the parties will "negotiate in good faith," that can be more than polite language. In some jurisdictions (Delaware is the one deal people talk about most because so many entities are formed there), a well-drafted "good faith negotiation" obligation can create a real duty: not a duty to close, but a duty not to sabotage the process or treat negotiations as theater while you shop for alternatives.

So the right mindset is blunt: *every sentence in the LOI either does nothing or can be used against you later.* The job is to know which is which and to draft accordingly.

What makes a term ``binding" in practice

A binding LOI term usually has three features, and you can spot them without a law degree.

First, it's operational. It tells someone to *do* something (or stop doing something) immediately: don't solicit, don't disclose, pay your own expenses, return information, choose a governing law. Second, it's definite enough to be policed: the clause can be violated in a way that a judge can understand. Third, it's valuable even if the deal dies. Confidentiality matters if there is no deal. Exclusivity matters if there is no deal. Expense allocation matters if there is no deal. Those clauses aren't dependent on closing; they're about surviving the breakup.

By contrast, the non-binding terms are usually aspirational and forward-looking. They describe a transaction the parties *hope* to sign later, subject to a pile of future work and future approvals. They can still be useful—they anchor drafting and reduce surprise—but they're not typically written in a way that a court can "force" the parties to perform.

Exclusivity (No-Shop): the golden handcuffs, and why buyers crave them

No-shop clauses feel rude because they are. They are the buyer asking the seller to stop flirting while the buyer spends real money getting comfortable. That money is not just legal fees. It's accounting fees (quality of earnings), IT diligence, tax structuring, environmental reviews, lender diligence, internal time, and the opportunity cost of focus. Buyers want exclusivity because diligence is expensive and because diligence is also revealing: the buyer will learn things about the business that, if leaked, could become a weapon in someone else's hands.

From the seller's perspective, granting exclusivity is like taking the business off the market and handing the only serious buyer the right to slow down. Sellers fear two things: time and leverage loss. Time kills deals by letting doubts multiply. Leverage loss happens because once the seller stops talking to others, the buyer can try to improve terms in a private room where there is no competitive tension.

That's why exclusivity is negotiated as hard as price in many real processes. The "non-binding economics" may be a handshake. Exclusivity is the part where someone actually gives something up.

The no-shop clause is not one promise; it is a bundle of promises

Bad exclusivity drafting treats no-shop like a single prohibition: "Seller shall not solicit other offers." That's the kindergarten version. Real no-shops are bundles of separate behaviors, each with its own loopholes if you don't close them.

Solicit is obvious: don't go fishing for other buyers. But what about *respond*—if someone calls out of the blue, can the seller take the meeting? What about *discuss*—can management keep chatting under the pretense of "industry networking"? What about *furnish information*—can the seller give a new buyer a CIM, updated financials, or data room access? What about *receive* proposals—can they even accept an inbound email that says "we'd pay more"? What about *encourage*—can they hint to a third party that now might be a good time to bid?

The answer is not "always forbid everything." The answer is: decide what you actually need, then draft it so the other side can't pretend it meant something else. Buyers want a clause that prevents the seller from creating a shadow auction. Sellers want a clause that doesn't freeze ordinary business communications or make them liable for someone else's inbound curiosity.

Then there's the "who is covered" problem. Is the restriction on the company only? Or does it cover "Representatives"—board members, officers, employees, advisers, investment bankers? If the LOI doesn't cover representatives, it can be exclusive in name while the banker runs a parallel process in practice. If it does cover representatives, sellers will want to ensure it doesn't create unreasonable policing obligations ("we can't control what someone says at a conference"). Precision matters because exclusivity disputes often turn on facts that sound petty until they cost eight figures.

Duration: the calendar is part of the price

Exclusivity is negotiated in days for a reason. A 30-day no-shop feels like a sprint. A 60-day no-shop feels like a semester. A 90-day no-shop can feel like a hostage situation, depending on the business and the buyer.

Buyers often argue for a longer period by pointing to the work required: third-party diligence timelines, lender processes, and the simple reality that drafting definitive agreements takes time. Sellers argue for shorter periods because they know that every day of exclusivity reduces their alternatives and increases the odds the buyer tries to renegotiate in private.

A practical compromise shows up frequently: a base exclusivity period plus an extension if certain milestones are met (for example, if the buyer delivers a near-final draft of the definitive agreement by day X, or if financing commitments are delivered by day Y). Milestone-based extensions are not just cute drafting; they are a way to align incentives. If the buyer wants the seller off the market, the buyer should move.

Fiduciary outs: not automatic, and not always appropriate

People throw around "fiduciary out" like it's a standard checkbox. In reality, whether it belongs in an LOI depends heavily on the seller and governance context. A fiduciary out is a carve-out that allows a board (or a governing body) to consider a superior proposal even during exclusivity if failing to do so would breach fiduciary duties.

The critical point is simple: you don't get a fiduciary out just because you believe you should have one. If the seller wants the ability to entertain superior proposals, it needs to be negotiated and written.

Buyers hate vague fiduciary outs because they can swallow exclusivity whole. Sellers need them when governance duties are real and the risk of being sued by their own owners is not theoretical.

Even where a fiduciary out exists, it is often structured with guardrails: notice requirements, a matching right (the buyer gets a chance to improve terms), and restrictions on what "discussions" can occur. The drafting is delicate because it's trying to reconcile two truths that do not naturally coexist: "we will negotiate exclusively" and "we reserve the right to stop negotiating exclusively."

After the no-shop: here's where the chart belongs

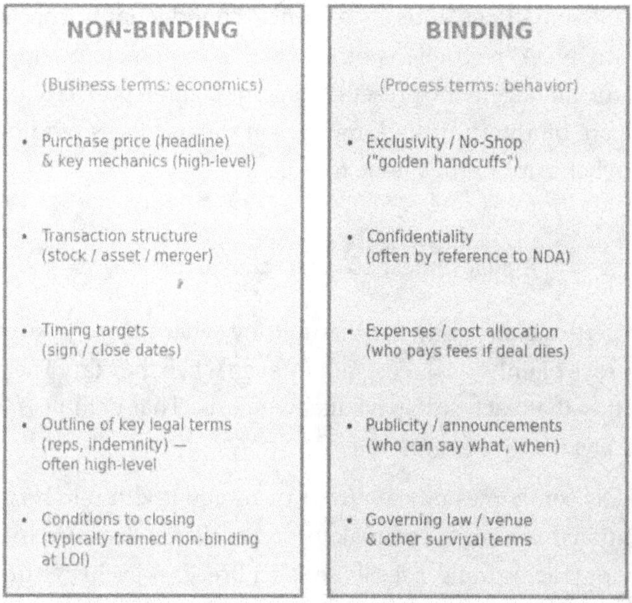

Practical takeaway: "Non-binding" usually means you can walk away from the deal — but you may not be free to shop, disclose, or stall while you do it.

Confidentiality: the clause that quietly does most of the work

Confidentiality is often treated as "already handled" because an NDA was signed early. Then the LOI references that NDA and everyone moves on. That's fine until the deal wobbles and the question becomes: what can the buyer do with what it learned?

Confidentiality is binding because it protects the seller even if nothing closes. It also protects the buyer in a subtler way: it forces a disciplined flow of information, which reduces the risk that diligence turns into uncontrolled rumor. For sellers, confidentiality is also about leverage. If information leaks during an auction, the seller's bargaining position can weaken with employees, customers, and competitors. For buyers, confidentiality breaches can be reputationally toxic, especially for sponsors and strategics who want to be welcomed into future processes.

The LOI sometimes adds incremental confidentiality concepts beyond the NDA: restrictions on contacting customers or employees, rules for management presentations, or tighter publicity controls. Those are binding for the same reason exclusivity is binding: they govern behavior in the messy middle.

Expenses: the most honest clause in the document

An expenses clause is the LOI admitting what everyone knows but rarely says plainly: deals cost money even when they fail. The market default is that each side pays its own costs. That default is simple, and that's why it's common.

But LOIs sometimes deviate from the default with reimbursement obligations, caps, or even break fees/cancellation fees tied to certain behavior (for example, a fee if the seller breaches exclusivity or if the buyer walks away after a defined milestone). These mechanisms are not about squeezing the other side for dinner money. They are about creating consequences when one party wants the other to incur significant costs early.

Expense reimbursement can also be a proxy for confidence. A buyer who asks for reimbursement if the seller breaches exclusivity is saying, "We're about to spend real money, and we need you to take exclusivity seriously." A seller who asks for a reverse break fee (a fee payable by the buyer in certain failure scenarios) is saying, "We're taking ourselves off the market, and we want protection against you using us as a free option." These are heavy asks, and they're not "standard" everywhere, but they exist because the underlying incentives are real.

Good-faith negotiation: the clause that sounds nice and can bite hard

"Good faith" language feels like manners. In the LOI context, it can be a commitment with teeth. The danger is not that it forces you to close; the danger is that it constrains how you can behave while deciding not to close.

A buyer who signs an LOI with a good-faith covenant and then treats diligence as a slow-motion fishing expedition may be handing the seller an argument that the buyer never intended to reach agreement. A seller who signs and then uses exclusivity as cover to stir up a new bidder may be handing the buyer an argument that the seller sabotaged the process. The clause invites a later inquiry into intent and behavior, which is exactly the sort of inquiry deal teams hate because it's expensive, fact-heavy, and hard to predict.

If parties want a "keep moving" obligation without opening the door to ambiguous good-faith litigation, they often do better with concrete process commitments: timelines, deliverables, weekly check-ins, a first draft date for definitive agreements, data room completion dates, diligence access commitments. Those are easier to measure than "good faith" and often accomplish the same practical goal: no one gets to quietly stop trying while still enjoying the benefits of exclusivity.

Drafting exclusivity like a grown-up: scope, conduct, and consequences

A no-shop clause earns its keep only if it answers uncomfortable questions.

What exactly is prohibited? Not just "solicit," but also "initiate," "encourage," "facilitate," "discuss," and "furnish information," if that is what the buyer truly needs. What about inbound proposals? If a seller can entertain them freely, exclusivity becomes performative. If the seller must shut them down immediately, the seller may worry about breaching duties or alienating important relationships. A middle ground sometimes appears: the seller can receive an unsolicited proposal, but must promptly notify the buyer and cannot engage in substantive discussions without triggering defined steps (and possibly ending exclusivity).

Who is bound? The company, its owners, its board, and its representatives. If advisers aren't covered, the clause can be circumvented with a phone call that "wasn't from the company." If employees aren't covered, informal discussions can begin under the guise of "networking." If owners aren't covered, side conversations can be justified as "personal."

What happens if it's breached? This is where exclusivity becomes more than a moral statement. Remedies matter. Many buyers push for injunctive relief language (the right to go to court to stop a breach, because money damages may not be adequate). Sellers may resist, but the logic is intuitive: if exclusivity is the buyer's bargain, the buyer wants the ability to protect it in real time, not collect a check after the deal is gone.

And finally: what is the buyer committing in exchange? Exclusivity without buyer momentum is where sellers get burned. Sellers can and should ask for a real timetable and a real diligence plan. Exclusivity is not a gift; it's a trade.

The behavioral trap: ``We're non-binding" can create sloppy conduct

The most common LOI failure mode is not a drafting error; it's a conduct error. People treat "non-binding" as permission to be casual. They forward diligence materials too widely. They talk too freely. They miss timelines. They "keep options open" in ways that violate exclusivity. They assume no one will enforce anything because "it's just an LOI."

But LOIs are often enforced precisely because they govern the stuff that happens *before* the definitive agreement. If a party breaches confidentiality or exclusivity, the harm happens immediately: information leaks, bidders appear, leverage shifts. That's why courts are more willing to treat these provisions seriously. The damage is not hypothetical; it is the entire point of the clause.

So the right discipline is to treat binding provisions like a miniature contract that starts operating the day it's signed. The non-binding provisions are the sketch of a future transaction. Confusing those categories is how deal teams accidentally create real liability while thinking they're still in the safe, early, "just talking" phase.

A clean way to think about it

The LOI is two documents stapled together. One is a draft of a future bargain. The other is a set of rules for how the parties behave while they decide whether that future bargain deserves to exist.

Exclusivity is the rule everyone remembers because it's the one that takes away options. Confidentiality is the rule everyone forgets until it's violated. Expenses are the rule that becomes important the second someone starts counting sunk costs. And good-faith language is the rule that can turn a breakup into a legal story about who acted like an adult.

A well-crafted LOI doesn't just say "binding" and "non-binding." It draws the boundary in thick ink, then writes the binding terms in a way that makes compliance easy, breaches obvious, and incentives aligned. That's not legal perfectionism. It's what keeps a deal process from turning into a referendum on what someone "meant."

2.4 Conditions and Deal Contingencies

Every buyer wants an escape hatch. Every seller wants to believe there isn't one. Conditions are where those two desires meet, shake hands, and immediately check each other's pockets.

A "condition" is just a requirement that must be satisfied before the parties are obligated to close. It sounds technical, but it's really about storytelling. The buyer is trying to say: "I'm serious, but I'm not reckless." The seller is trying to hear: "You're not about to spend 60 days turning my company inside out and then ghost me." The LOI has to bridge that gap without falling into two common traps: conditions that are so broad the offer looks fake, and conditions that are so narrow the buyer is effectively underwriting unknown risks.

The subtlety is that the LOI is not the purchase agreement. You are not drafting the definitive closing conditions yet. You're drafting the *expectations* that will govern the definitive drafting, and you're doing it in a way that affects leverage immediately. A well-crafted contingency section doesn't read like a list of excuses. It reads like a plan: what needs to be proven, what needs to be obtained, and what needs to happen in what order.

The seller's fear: ``conditions" as camouflage

Sellers don't mind conditions in principle. They mind conditions that function like a silent auction strategy: tie up the company under exclusivity, learn everything you can, then decide later whether you "like what you see." If the seller has lived through a busted process, they can smell this a mile away. The odor is usually the same phrase repeated in different disguises: "to the buyer's satisfaction."

"Satisfactory due diligence." "Satisfactory financing." "No new material information." Those are not conditions; they're vibes. They can be legitimate in narrow contexts, but in an LOI they often communicate a single message: *we reserve the right to change our mind for any reason*. If you're a seller reading that, you should worry. If you're a buyer writing that, you should realize you're buying yourself flexibility by selling the seller a reason to distrust you.

A buyer can protect itself without spooking the seller by doing something surprisingly rare in LOIs: naming the real risks out loud. Not with melodrama, but with specificity. Sellers can accept "we need to get comfortable with customer retention by cohort and gross margin quality" because that sounds like diligence, not opportunism. They react badly to "we need to be satisfied with everything," because that sounds like a free option dressed up as prudence.

The buyer's fear: committing to a price while still guessing

The buyer's problem is simple and brutal. At LOI stage, the buyer is being asked to put a number on a business while still missing information that could move that number meaningfully. Customer contracts might not be fully reviewed. Net working capital may not be normalized. Revenue recognition might have quirks. There could be unpaid sales taxes hiding in a state you've never visited. The company might have one customer who accounts for 35% of revenue and one key engineer who accounts for 80% of the product roadmap.

Conditions are how buyers keep those unknowns from turning into losses. But there's a big difference between protecting against *unknown unknowns* and preserving the right to renegotiate *known knowns*. Sellers can live with the first. They hate the second.

So the best LOI contingencies do two things at once:

- they ring-fence the handful of issues that could genuinely be deal-breakers, and

- they show the seller a credible path to removing those conditions quickly.

That second part is what keeps conditions from looking like a stall tactic. A condition without a plan is just a threat in paragraph form.

Confirmatory diligence: the most abused phrase in the process

"Confirmatory diligence" is supposed to mean: you've already done enough work to believe the deal is real; what remains is verification, not discovery. In a fast auction, confirmatory diligence might be all you get. In a bilateral deal, it's often the promised future state: once you sign the LOI, the buyer says it will be confirmatory.

The problem is that the phrase is routinely used as a social lubricant. Buyers use it to reassure sellers. Sellers use it to pressure buyers into moving faster. Neither side defines it. Then everyone is surprised when it turns out the buyer still has major questions.

If you want "confirmatory diligence" to mean anything, it has to be tied to a list of focus areas and an access commitment. Otherwise, it becomes a blank check: the buyer can always claim something wasn't "confirmed," and the seller can always claim the buyer is expanding scope. The LOI is the right place to reduce that ambiguity, because once exclusivity is signed, ambiguity turns into leverage—and someone will use it.

The three big condition buckets (and why they feel different to sellers)

Most LOI conditions fit into three broad categories: diligence, financing, and approvals/definitive documents. They're not equal in seller psychology.

Diligence conditions are expected, but sellers want them to be bounded. Financing conditions are the most emotionally loaded, because sellers hear them as "we may not be able to pay." Approvals and definitive documents are often treated as boilerplate, but they hide a real risk: either side can drag its feet on drafting to create leverage or to run out the clock.

A smart buyer recognizes the emotional hierarchy and drafts accordingly. You can get more protection by being more precise, not by being more expansive.

Diligence outs: protect against lies, not against learning

There is a clean moral way to draft a diligence condition: "We need to verify that what we were told is true." Sellers may not love it, but they can respect it.

There is also a messy way: "We need to see if we feel good after learning more." That is where sellers start to imagine being trapped in exclusivity with someone shopping for excuses.

The highest-quality diligence contingencies are anchored to objective concepts. For example:

- *Financial verification*: quality of earnings (QoE), revenue recognition, margin sustainability, customer concentration, backlog integrity.
- *Legal verification*: ownership of IP, key contracts, change-of-control/assignment restrictions, litigation exposure.
- *Operational verification*: inventory existence/obsolescence, cybersecurity posture, compliance regimes if regulated.

You don't need to list every workstream, but you do want to signal that diligence is about validation, not about re-shopping the deal you already priced.

One surprisingly effective phrasing technique is to connect diligence to the information set already provided: "subject to confirmatory diligence consistent with the information provided to date." That reassures the seller that the buyer is not planning a total reinvention of the deal while still leaving room for the buyer to react if the information provided was incomplete or misleading.

Financing contingency: the clause sellers hate (and buyers often need)

If a buyer needs debt financing, a financing contingency can be the difference between prudence and disaster. Without it, the buyer risks being obligated to close even if lenders refuse to fund, which can trigger default, reputational damage, and sometimes litigation. With it, the seller risks spending months under exclusivity only to be told, at the end, "the banks said no."

This is why financing contingencies are one of the sharpest negotiation points in the LOI. In competitive processes, sellers often prefer buyers who can waive financing conditions or present committed financing early. In bilateral negotiations, sellers may tolerate financing contingencies but will usually demand comfort that the buyer is financeable.

There are ways to draft financing protections without sounding like you might not have the money:

- *Commitment-based language*: instead of "subject to financing," the buyer indicates it will seek debt financing promptly and may provide evidence of lender engagement (term sheets, committed letters, or at least a credible financing plan).
- *Limited contingency scope*: financing is conditioned on obtaining financing on terms consistent with a described structure, rather than "any financing acceptable to buyer."
- *Buyer covenant to use efforts*: the buyer commits to use commercially reasonable efforts (or "reasonable best efforts," depending on the deal) to obtain financing. This matters because it turns the contingency from a free option into an obligation to try.

The seller still bears risk, but the risk becomes about market reality rather than buyer whim.

The financing mismatch problem: when your lender has an out you don't

Here's a practical landmine that shows up in real deals: the lender's funding conditions can be broader than the buyer's closing obligations under the acquisition agreement. That mismatch creates a terrifying situation for a buyer: you are obligated to close, but your lender can refuse to fund.

At LOI stage, you're not negotiating the full debt commitment package, but you can set expectations that help prevent the mismatch later. The buyer should be thinking: if the purchase agreement says I must close unless there's a material adverse effect (or unless specific conditions fail), will the lender have a "MAC" (material adverse change) out that is broader? Will the lender require "satisfactory diligence" in a way that allows subjective refusal? Will the lender impose conditions about projections, customer interviews, or collateral appraisals that could become choke points?

Even a short LOI sentence acknowledging this alignment can be valuable: financing efforts will be pursued on terms that are intended to be consistent with the closing conditions and timeline contemplated by the definitive agreement. It won't solve everything, but it signals to the seller that the buyer understands the problem and is not setting up a silent escape hatch through the debt documents.

Regulatory approvals and third-party consents: the conditions that look boring until they aren't

Some conditions are "boring" only because people underestimate them. Third-party consents—especially in asset deals where contracts don't transfer automatically—can turn a straightforward acquisition into a months-long consent hunt. Regulatory approvals can be routine or can swallow the deal.

In an LOI, you usually don't need to enumerate every potential consent. But it's often worth identifying any obvious gating consents early: a landlord consent for a critical facility, a key customer contract with strict assignment language, a software license that can't be transferred without permission. When these are ignored at LOI stage, they return later as either (a) a timing problem ("we can't close by the date we promised") or (b) an economic problem ("we need a price adjustment because we can't secure consent from the biggest customer").

A seller will often be more willing to accept a condition if the buyer ties it to a plan: which consents, what approach, what timeline, and who is responsible. Conditions feel less like excuses when they come with logistics.

Employment and retention conditions: when ``people risk" is actually deal risk

LOIs sometimes include conditions related to key employees: employment agreements, non-competes (where enforceable), retention bonuses, or rollover equity arrangements for management. Sellers can find these conditions intrusive, especially if they imply that the buyer is negotiating with management while the owners are still negotiating with the buyer.

But in many businesses, people are the product. If the company's value depends on three engineers, a sales leader, or a founder who holds customer relationships in their phone, then "key employee arrangements" are not fluff. They are risk management.

The drafting challenge is tone. A blunt condition ("key employees must sign employment agreements acceptable to buyer") reads like the buyer wants veto power over the human beings who make the company run. A more workable approach frames it as alignment: the parties will use good-faith efforts to agree on retention arrangements for identified key personnel, consistent with market terms. It's still a condition in substance, but it sounds like a joint project rather than a demand.

Material adverse change (MAC) concepts: the early whisper of a later fight

Even if an LOI doesn't include a full "MAC" closing condition (and most don't go into that detail), it's common to include at least a notice concept: the seller must inform the buyer promptly of any material adverse developments prior to signing or closing. That can sound innocuous, like common sense. It's also a signal that the buyer is thinking about what happens if the business deteriorates during exclusivity.

Sellers worry about MAC language because it can become a lever for renegotiation if performance softens. Buyers worry about the opposite: committing to a price and timeline while the business changes in ways that would have affected the price.

The LOI sweet spot is often to keep MAC language high-level and process-oriented: prompt notice, continued ordinary-course operation, no extraordinary actions without consent. That sets the stage for the definitive covenant package without turning the LOI into a pre-fight about what constitutes a "material adverse effect."

Definitive documents as a ``condition'': boilerplate that can still be weaponized

"Subject to negotiation and execution of definitive agreements" is standard LOI language. It's also the clause that can be used to justify bad behavior: endless markups, slow drafting, strategic delay.

That's why the best LOIs pair the definitive-agreement condition with a process: who drafts first, when the first draft is due, when comments are due, and what the target signing date is. These don't have to be binding commitments, but they create an expectation that can be enforced socially, which is often how deals actually move.

In auctions, the seller often controls this by providing a near-final purchase agreement early and asking bidders to mark it. In bilateral deals, the buyer may need to push for drafting discipline explicitly, because without a process manager, "we'll get you a draft soon" can stretch into a month.

A word on ``satisfaction'' standards: the difference between protection and provocation

"Satisfactory to buyer" can be appropriate when the item is inherently subjective, but it is overused. Sellers see it as an invitation to fabricate dissatisfaction. Buyers see it as necessary flexibility.

One way to reduce friction is to swap subjective "satisfaction" language for objective or semi-objective standards:

- "in all material respects consistent with" the information provided,
- "no material discrepancies identified in" specified diligence workstreams,
- "no material changes in" specified metrics, coupled with a defined baseline.

These aren't perfect either, but they signal that the buyer is not seeking a mood-based out.

And if the buyer truly needs a subjective standard, it can be softened by pairing it with an efforts covenant and a tight timeline. "Satisfactory diligence" sounds less like a trap when the buyer commits to complete diligence by a date certain and to provide a written list of open issues.

Contingencies as relationship management: the art of sounding serious

The best contingency drafting is oddly empathetic. It acknowledges the seller's risk: time off market, operational distraction, confidentiality exposure. It also acknowledges the buyer's risk: writing a big check for a business that hasn't been fully verified.

That empathy can be operationalized. If the buyer asks for a diligence condition, it should also ask for diligence access rights and agree to cadence: weekly diligence calls, a Q&A protocol, a commitment to keep requests reasonable, and a mechanism to resolve disputes quickly. If the buyer asks for a financing condition, it should commit to immediate lender engagement and to deliver evidence of progress. If the buyer asks for consent-related conditions, it should propose a consent plan.

Sellers don't get spooked by conditions. They get spooked by *unbounded conditions with no urgency*. Show urgency, and conditions look like prudence. Hide behind vagueness, and conditions look like a strategy to keep your options open while closing theirs.

The real purpose of LOI conditions: decide what is allowed to change later

Conditions are often described as "outs," but that framing is too cynical. The better framing is that conditions decide which facts are allowed to matter later.

If diligence is a condition, then new information can legitimately change the buyer's willingness to proceed. If financing is a condition, then the capital markets can legitimately veto the deal. If key employee arrangements are a condition, then people risk can legitimately stop closing. If consents are a condition, then counterparties can legitimately hold the keys.

That might sound obvious. It isn't. Many busted deals are not caused by new information; they are caused by disagreement over whether the new information was allowed to matter. A seller calls it re-trading. A buyer calls it risk adjustment. The LOI's contingency section is where that dispute is either prevented or preloaded.

Well-written LOI conditions don't kill momentum. They channel it. They take the buyer's uncertainty and turn it into a checklist with deadlines. And they take the seller's fear of being strung along and turn it into process commitments that force the buyer to keep moving or stop pretending.

Chapter 3

The Due Diligence Engine

Think of signing a Letter of Intent as agreeing to buy a house based on a curbside view. Due diligence is the moment you finally walk through the front door, open the closets, and check the foundation for cracks. It is not merely an audit; it is a strategic engine used to validate the investment thesis and map out the future. In this chapter, we move beyond the checklist mentality to show you how to run a coordinated investigation-covering money, law, and tax-and how to turn those discoveries into the protective shield known as disclosure schedules.

3.1 Planning and Managing Diligence

Due diligence has a reputation problem. People imagine a grim scavenger hunt through PDFs, where an army of associates hunts for missing signatures and expired leases like it's a competitive sport. In reality, diligence is closer to running a newsroom on a deadline: you have limited time, imperfect information, and the uncomfortable knowledge that whatever you miss will become *your* problem right after closing. The goal isn't to know everything. The goal is to find the things that can (a) change price, (b) change structure, (c) change the willingness to do the deal, or (d) become a post-close fire you'll be paying to put out.

The most dangerous version of diligence is the one that feels busy. That's when the request list grows like ivy—more folders, more documents, more late-night emails—while the real risks hide in plain sight. A well-run process is opinionated: it has a theory of what matters, a way to test that theory quickly, and a mechanism to escalate bad news without derailing the entire machine. You're not trying to *win* diligence. You're trying to turn uncertainty into decisions.

Start with a risk thesis, not a template

A template diligence list is like a hotel gym: technically useful, emotionally misleading. It makes you feel virtuous just by showing up. But the best diligence plans start with a blunt question: "Where could this business be lying to me—intentionally or accidentally?"

That question is not an accusation. It's realism. Businesses are messy. Management teams tell stories for a living. Accounting systems are imperfect. Contracts were written by people who have since left. Even honest companies can present a picture that's true in spirit and wrong in mechanics. Diligence is the discipline of translating a narrative ("we have sticky customers and recurring revenue") into testable claims ("what does churn look like by cohort, and do the contracts allow repricing?").

A practical way to build the risk thesis is to force the deal into a handful of "value drivers," then pre-assign a skeptic to each one. If the valuation depends on gross margin expansion, someone needs to interrogate pricing, discounting, freight, returns, warranty reserves, and labor assumptions like they're cross-examining a witness. If the valuation depends on a single customer, someone needs to read that contract like a breakup letter: what's the easiest way this relationship ends, and how quickly can it happen?

The risk thesis also keeps you honest about time. Diligence has a clock attached to it—exclusivity, financing deadlines, board meetings, and the seller's patience. If you treat every request as equally important, you guarantee that the important ones arrive last, when the only tool left is panic.

The request list should be a weapon, not a wish list

A request list is often treated as a polite suggestion: "When you get a chance, could you upload three years of customer invoices?" That tone is friendly, but it's strategically naïve. The request list is a control document. It is how you create accountability, sequence the work, and expose where the seller is slow, disorganized, or evasive.

A high-functioning request list has three traits.

First, it is prioritized by consequence. "High consequence" doesn't mean "interesting." It means "deal-moving." If you could discover a problem that changes enterprise value by 10%, it goes to the top even if it's boring. Working capital mechanics are boring; they also decide who writes a check to whom after closing. Sales tax exposure is boring; it can also be a seven-figure surprise that doesn't care about your purchase agreement's feelings.

Second, it is written so the seller can comply without guessing what you meant. "Provide revenue by customer" is ambiguous. "Provide monthly revenue by customer for the last 36 months, including invoice date, invoice amount, product/service category, and any credits or refunds applied" is annoying—but it is executable. Diligence slows down when requests are interpretive. The seller's team will respond with whatever is easiest, not what is useful, and you'll burn a week arguing about definitions.

Third, it is instrumented. Every request needs an owner, a date requested, a due date, a status, and a place for notes. The notes are not fluff; they become your institutional memory. "Seller says they don't track this" is an important fact. It means your integration plan just got more expensive, or your confidence in forecasts just got smaller, or both.

One subtle trick: separate "upload requests" from "analysis requests." Upload requests live in the VDR and can be tracked like inventory. Analysis requests—things like "reconcile management EBITDA to audited statements" or "explain margin decline in Q3"—

often require the seller's finance team to do work, not just drag files. Treat those as meetings and deliverables, not as line items buried in the document pile.

Triage: red flags, yellow flags, and the difference between bad and fatal

Not every ugly discovery is a deal-killer. Some are simply expensive. Some are embarrassing but containable. Some are a sign you're buying a risk factory with a revenue side hustle.

Triage is the muscle that separates a professional diligence process from a nervous one. Without triage, every finding triggers the same response: more questions, more documents, more meetings. That feels thorough until it becomes the reason the deal dies of exhaustion.

A workable triage framework is simple:

- *Red flags*: issues that can stop the deal, force a major restructure, or create open-ended liability that can't be priced with confidence.
- *Yellow flags*: issues that are real and painful but can be translated into price, terms, or a post-close plan.
- *Green-ish noise*: imperfections that are normal for the size and maturity of the company and don't change your underwriting.

The trick is not the labels; it's the discipline to attach an action to each one. A red flag triggers escalation and decision-making: do we walk, pause, or renegotiate? A yellow flag triggers a translation exercise: can we quantify it, or at least bound it, and then decide whether it becomes a price adjustment, an indemnity request, a covenant, or an integration priority? Green-ish noise gets documented and then ignored so the team can keep breathing.

A common failure mode is to confuse "surprising" with "material." For a first-time buyer, a messy cap table can feel like an earthquake.

For a seasoned deal team, it may be a Tuesday—unless it touches ownership of key assets or complicates tax structure. Surprise is an emotional signal; materiality is a financial and legal one.

The Virtual Data Room (VDR) is not storage---it's the factory floor

People talk about a VDR like it's a filing cabinet. That's the wrong mental model. A VDR is the factory floor where diligence is manufactured: inputs arrive, get processed by workstreams, and produce outputs that change deal terms. If the factory floor is chaotic, production slows down, defects increase, and everyone starts blaming each other for problems that are really just poor workflow.

The best VDRs feel boring in the same way a well-run warehouse feels boring. You can find things. The naming is consistent. There aren't six versions of the same file with timestamps that look like a cryptic puzzle. The folder structure reflects how people think about the business, not how someone happened to save documents on a desktop in 2019.

A few practical rules make an outsized difference:

Use a folder structure that mirrors diligence workstreams

If you organize the VDR like the seller's internal shared drive, you inherit their chaos. Organize it like the buyer's diligence plan: Financial, Tax, Legal, HR, Commercial, Operations, IT/Cyber, Environmental (if relevant), and so on. Within each, create predictable subfolders (e.g., "Financial / Monthly P&Ls," "Financial / Revenue detail," "Financial / Debt and leases," "Legal / Material contracts," "Legal / Litigation," "Tax / Returns," "Tax / Audits and notices"). Predictability is speed.

Enforce naming conventions that age well

A file named "Final_Final_v3.xlsx" is a cry for help. Require naming that survives context loss: date, topic, period, and version if needed (e.g., "2025-12 Customer_Revenue_Detail_Jan2023-Dec2025 v1.xlsx"). A year from now, someone will need to understand what you relied on, and the VDR will be the only witness that doesn't forget.

Track superseded files deliberately

Sellers often upload revised schedules. That's normal. The danger is when no one can tell what changed and why. When a revised file arrives, require a short change log in the Q&A notes or as a companion upload. "Updated AR aging to reflect credits issued after year-end" is a helpful sentence. "Updated" is not.

Treat Q&A as part of the record, not hallway chatter

Many VDRs have Q&A functionality for a reason: it captures the question, the answer, and the date. The worst alternative is a universe of side emails and informal calls where the most important clarifications live only in someone's memory. If the answer changes the model, it should be written somewhere durable.

Control permissions like you control leaks

Diligence is a confidentiality exercise as much as it is an analysis exercise. Not everyone needs access to everything. Customer contracts, employee data, security policies, and pricing files require tighter controls. A buyer that treats confidentiality casually creates a seller that responds by slowing down, redacting aggressively, or refusing to provide sensitive materials until the last possible moment.

Parallel workstreams: the art of moving fast without breaking the deal

Diligence runs on parallel tracks because the questions are different, the specialists are different, and the timelines are different. Financial diligence wants trial balances and revenue detail. Legal diligence wants contracts and corporate records. Operations wants plant tours and KPI dashboards. IT wants architecture diagrams and incident logs. Tax wants returns, notices, and entity structures. Each group has its own language and its own definition of "urgent."

Coordination is what prevents parallel tracks from becoming parallel universes.

One useful habit is to establish "workstream leads" who own two jobs: doing the work and translating the work. Translation matters because diligence findings rarely stay in their lane. A legal issue becomes a financial issue when it creates a liability reserve. A financial issue becomes a legal issue when it needs to be converted into a representation, a covenant, or a schedule disclosure. An IT weakness becomes a commercial issue when downtime threatens customer retention.

A weekly steering meeting is necessary but not sufficient. The real coordination happens in short, frequent touchpoints where leads surface blockers early: missing data, inconsistent reporting, a seller who insists a metric is "not tracked," a material contract that can't be located, or an explanation that changes every time you ask it.

The discipline is to keep these touchpoints crisp. Diligence meetings expand to fill the time available. If you allow that, you'll have long conversations about minor items because they're easier than confronting the hard ones. A good agenda is mostly exceptions: what's late, what's inconsistent, what's materially weird, and what requires a decision.

The hidden killer: reconciliation between different ``truths"

In many deals, the slowest part of diligence isn't a complex legal analysis or a sophisticated operational assessment. It's the awkward reconciliation of multiple versions of reality.

You will often see at least three "truths" about the company's financial performance:

- The story told in management reporting (often optimistic, sometimes adjusted).
- The story told in financial statements (more formal, sometimes lagging).
- The story told in tax returns (conservative by design, sometimes strategically so).

When these don't tie, the buyer's brain does what brains do: it assumes the worst. Sometimes that's unfair; often it's rational. Either way, a gap between narratives becomes risk, and risk becomes a discount, a delay, or a demand for protection.

A well-managed diligence process anticipates this. Early in the timeline, ask for a "source of truth map": what systems feed the P&L, how revenue is recognized in practice, how the general ledger is maintained, and what adjustments management makes when presenting "EBITDA." You're not asking for perfection. You're asking for a coherent explanation that stays coherent.

When the explanation changes every week, it's rarely because the truth is complicated. It's because no one is steering the story. That's a diligence finding in itself: it tells you something about controls, forecasting reliability, and how painful integration will be.

Diligence is an interview of the management team---whether you admit it or not

Documents are necessary. They are also a form of performance art. A seller can curate a VDR the way a homeowner stages a house: hide the clutter, repaint the ugly wall, and keep the lights flattering. That's not immoral; it's human. The counterbalance is direct interaction with the people who run the business.

Every management presentation contains two layers. The first is the content: market, strategy, KPIs, challenges. The second is the signal: how quickly they answer, whether they acknowledge uncertainty, whether they know their numbers, and whether they treat bad news as a problem to solve or a fact to conceal.

Diligence planning should explicitly allocate time for management Q&A by topic, not as a single all-hands event where hard questions get lost. Finance questions should be asked with the controller (or whoever actually closes the books), not only with the CFO who may be more storyteller than mechanic. Operations questions should involve the plant manager or service leader who lives with the bottlenecks. Customer retention questions should include whoever owns renewals, not just whoever owns the slide deck.

Good diligence makes these conversations easier by grounding them in documents. "Walk me through your churn" is abstract. "These top 20 customers show declining spend in the last two quarters—what happened?" is concrete, and concrete questions produce useful answers.

The cadence: front-load the deal-breakers, then deepen

Diligence should feel like a funnel. Early on, you want fast tests that can invalidate the deal thesis quickly. Later, you want deeper work that improves pricing precision, sharpens integration plans, and prepares drafting.

Early tests might include:

- Customer concentration and contract transferability.
- Revenue quality (recurring vs. project-based; refunds; unusual quarter-end spikes).
- Gross margin drivers and volatility.
- Debt, liens, and off-balance-sheet obligations (leases, guarantees).
- Material litigation, regulatory exposure, and compliance posture.
- Ownership of core IP and any "who actually owns this?" risk.

If these look acceptable, you earn the right to go deeper: cohort analyses, unit economics by product line, detailed working capital behavior, systems and cyber assessments, and operational capacity constraints.

This sequencing isn't about laziness. It's about respecting that diligence has an opportunity cost. Every hour spent on low-probability issues early is an hour not spent confirming the assumptions that justify the purchase price.

Escalation without drama: how to surface bad news and keep moving

Bad news in diligence is inevitable. The only question is whether it arrives early enough to be useful. Escalation is the mechanism that converts "someone noticed something" into "the deal team makes a decision."

A clean escalation protocol has three elements:

Define materiality thresholds in advance

If a workstream lead finds a potential $200k exposure, does that matter? Maybe, maybe not. If they find a $5 million exposure, it almost certainly does. The thresholds depend on deal size and the buyer's risk tolerance, but the existence of thresholds prevents two bad outcomes: junior team members hiding issues because they don't want to be wrong, and senior team members being spammed with noise.

Use "issue memos" that force clarity

An issue memo is not a novella. It's a one-page discipline tool:

- What is the issue (in plain language)?
- What is the worst-case exposure?
- What is the likely exposure?
- What evidence supports that range?
- What do we need to decide or request next?

This format forces the workstream to move from discovery to judgment. It also gives the deal lead something they can actually use in a call with an investment committee, lender, or board.

Separate "facts" from "positions"

Sellers will negotiate inside diligence. That's normal. The buyer says, "This looks like debt-like." The seller says, "That's just normal operations." Those are positions. They matter, but they can cloud the facts. Good escalation keeps them distinct: first establish what is true (the contract terms, the accounting treatment, the historical pattern), then argue about interpretation.

When you mix facts and positions too early, meetings turn into debates without evidence, and you lose time—the one resource you can't buy.

Keeping the deal moving: deadlines, friction, and the psychology of momentum

Deals die more often from fatigue than from a single fatal flaw. People underestimate how much stamina a transaction consumes: management distraction, advisor coordination, document churn, and the constant hum of uncertainty. A diligence process that drags becomes its own red flag. If it takes weeks to obtain basic schedules, what will integration look like? If answers arrive only after repeated prompting, what else is being avoided?

Momentum is created by small, visible progress:

- A VDR index that becomes more complete each day.
- A request list with closing percentages that climb.
- A weekly "open issues" log that shrinks, not grows.
- A calendar of management sessions that actually happens.

These are not cosmetic metrics. They affect behavior. Sellers respond faster when they see you're organized and decisive. Buyers stay aligned when they see issues being triaged and resolved. Advisors stop freelancing when they see a clear process they can plug into.

Friction is also information. If the seller consistently misses deadlines, gives partial responses, or claims not to have standard documents, treat that as a finding. Sometimes it's a resource constraint; sometimes it's disorganization; sometimes it's concealment. Either way, it has economic consequences: it increases the cost of diligence, increases integration burden, and decreases confidence in forecasts. Confidence is a pricing input even when no one writes it down.

How diligence planning protects your negotiating leverage

One of the quiet ironies of M&A is that the buyer often loses leverage precisely when they learn the most. Early on, before diligence, the buyer can credibly threaten to walk because they haven't invested much time or money. Later, after spending heavily on advisors and internal effort, the buyer may feel trapped by sunk costs. Sellers sense this. The negotiation shifts from "convince me" to "how much do you want this deal?"

Diligence planning is leverage protection. When you front-load high-risk questions, you discover deal-breakers while you still have the emotional ability to walk. When you keep a tight issue log, you prevent the seller from reframing a known problem as a late surprise. When you document Q&A, you avoid the slow erosion of truth where someone later says, "We never said that."

Most importantly, when you run diligence like a process rather than a scramble, you create a credible posture: you are willing to move fast, but you are not willing to be confused. Confusion is where bad deals hide.

The ``definition trap": diligence findings don't matter if they can't be operationalized

A diligence team can identify dozens of issues and still fail the deal if those issues aren't translated into something actionable. Actionable might mean a price change, a structure change, a contractual protection, or an integration plan. But "we noticed this" is not a deal term.

This is where planning matters more than intellect. If the team knows, from the beginning, that diligence outputs must be convertible into decision-ready language, they'll ask better questions. They'll quantify where possible. They'll capture contract terms precisely. They'll distinguish between "we think" and "we can prove."

A concrete example: suppose you discover that vacation accruals are under-recorded. That's not just an accounting observation. It can become (a) a purchase price adjustment if it changes working capital or debt-like items, (b) a specific pre-close action the seller must take (record the accrual), or (c) a post-close cleanup cost you accept with eyes open. The same fact can lead to three different deal outcomes depending on how it's framed and how quickly it's escalated.

A practical operating model: run diligence like a production system

A buyer who treats diligence as a production system tends to win—not because they're aggressive, but because they're predictable. Predictability is comforting in a transaction where everything else feels negotiable.

A simple operating model looks like this:

- *Daily*: request list updates, VDR uploads reviewed, urgent Q&A captured.
- *Twice weekly*: workstream lead standups focused on blockers and material issues.
- *Weekly*: steering meeting to review red/yellow flags, decide actions, and update timeline.
- *Continuous*: issue log maintained with owners and next steps; escalation protocol enforced.

The goal is not bureaucracy. The goal is rhythm. Rhythm reduces the cognitive load on everyone involved and makes it harder for important items to be "forgotten" in the noise.

When the rhythm is working, you can feel it. The VDR becomes a living workspace rather than a graveyard. Questions get sharper over time rather than more scattered. Findings arrive with suggested actions, not just anxiety. And the deal team spends its energy on judgment—what to do about what you found—instead of spending its energy on archaeology.

3.2 Financial Analysis and Quality of Earnings

Financial diligence is where a deal stops being a story and turns into arithmetic. Not the clean arithmetic of audited statements—messier, more revealing math: which customers actually pay on time, how often "one-time" expenses show up, whether margins are real or just a temporary truce between pricing and costs. If legal diligence is about what you *can* be forced to live with, financial diligence is about what you *will* be forced to live with—because cash has a way of making arguments short.

The center of gravity is usually a Quality of Earnings report—"QoE" in deal shorthand. A QoE is not an audit, and it's not an accusation. It's a translation exercise. It takes the company's reported results and asks a blunt buyer question: *How much profit is repeatable, and how much is theater?* The reason this matters is simple: valuation is often a multiple of EBITDA, and EBITDA is one of the easiest numbers on earth to misunderstand without anyone technically lying.

EBITDA: the number everyone quotes and nobody agrees on

EBITDA stands for Earnings Before Interest, Taxes, Depreciation, and Amortization. In plain English: "profit" before financing choices (interest), jurisdiction and structure (taxes), and certain non-cash accounting charges (depreciation and amortization). It's popular because it's quick, comparable, and feels like operating performance. It's also dangerous because it's incomplete.

EBITDA ignores working capital—the cash tied up in receivables, inventory, and payables. A business can show healthy EBITDA while consuming cash like a bonfire if customers pay slowly or inventory balloons. EBITDA also ignores capex (capital expenditures) and assumes depreciation is a reasonable stand-in for what the business needs to reinvest. For an asset-light services firm, that might be fine. For a manufacturer with aging equipment, it can be fantasy.

Despite the flaws, EBITDA remains the deal's common language. The QoE doesn't try to overthrow that reality. It tries to make the EBITDA you're using feel less like a slogan and more like something you'd defend in front of your investment committee with a straight face.

The QoE mindset: recurring versus convenient

A good QoE is built on a single obsession: sustainability. It's less interested in whether the company followed every accounting rule perfectly and more interested in whether next year looks like last year for reasons that will still be true after you own it.

That sounds abstract until you watch how it plays out. Buyers don't pay a premium multiple for a profit spike caused by a one-time vendor rebate. They pay for durable gross margin, repeatable renewal behavior, and cost structures that don't depend on a founder's superhuman tolerance for chaos. The QoE pokes at anything that looks unusually helpful to the current period.

The hard part is that "unusual" is not the same as "non-recurring." A big legal settlement is usually non-recurring. A "temporary" underinvestment in maintenance can recur for years—until the building, the servers, or the customer relationships finally revolt. The QoE's job is to keep asking: *Is this a real improvement, or just a bill that hasn't arrived yet?*

Add-backs: the place where optimism goes to get dressed up

Most QoE debates orbit one word: *add-backs*. An add-back is an adjustment that increases EBITDA by removing an expense (or reversing a reduction) that the seller claims is non-recurring, non-operating, or otherwise not representative of future performance. Conceptually, it's reasonable. Practically, it's where the seller's best arguments—and worst habits—tend to gather.

Legitimate add-backs exist. Owner-run businesses often run personal expenses through the P&L: family cell phones, a car lease that's more lifestyle than business necessity, maybe even the occasional travel expense that looks suspiciously like a beach. Those costs *are* real expenses, but they may not be expenses you'll have after closing. Adjusting for them can be fair.

Add-backs get aggressive when they start to remove the cost of being a functioning company. "Non-recurring" becomes a costume that any expense can wear if it wants to sneak out of the valuation. The classic offenders are:

- "One-time" professional fees that appear every year under a different label.
- "Temporary" labor or contractor costs used to patch chronic staffing problems.
- Expenses deferred by not replacing broken systems, not paying bonuses, or stretching vendors.
- Revenue boosts that came from unsustainable discounting or end-of-quarter heroics.

A disciplined buyer treats add-backs like a courtroom. The seller isn't guilty—but the evidence matters. Each add-back needs a narrative *and* support: invoices, contracts, board minutes, payroll detail, or at least a coherent explanation that doesn't change depending on who's asking.

The most buyer-friendly way to think about add-backs is to sort them into two buckets:

- *"Will not exist after close"*: expenses tied to ownership (above-market owner salary, personal expenses) or truly one-time events (a discontinued facility, a one-off lawsuit settlement).
- *"Should not exist, but might"*: inefficiencies management claims are fixable, such as excess headcount, unusual scrap, or "temporary" freight premiums.

The first bucket can be underwritten relatively cleanly. The second bucket is where deals quietly overpay. If the improvement depends on execution, it belongs in your value-creation plan, not automatically inside your purchase price multiple.

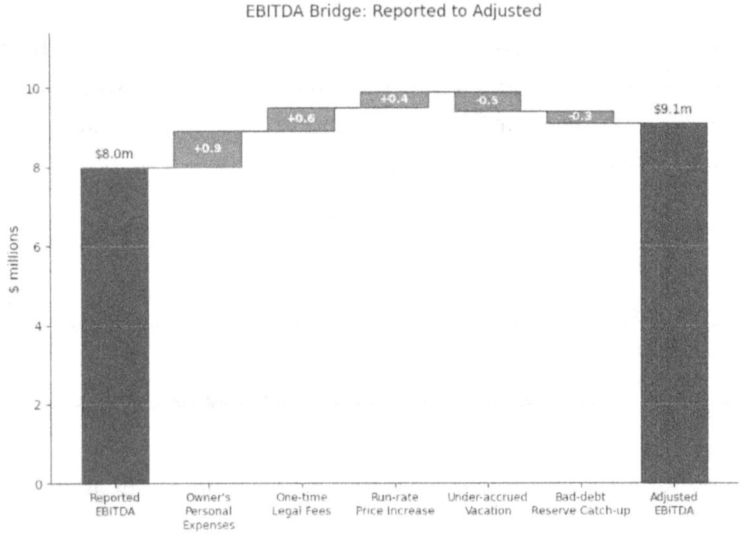

Negative adjustments: when EBITDA was flattered by omission

Buyers sometimes focus so hard on seller add-backs that they forget the other half of the bridge: adjustments that *reduce* EBITDA because the reported number benefited from under-accruals, optimistic estimates, or timing games.

These negative adjustments are emotionally harder because they feel like taking something away, and they often involve judgment. A few repeat characters show up across deals:

- *Under-accrued liabilities*: vacation, bonuses, commissions, warranty obligations, returns, rebates—anything that is economically real but can be delayed in accounting.

- *Bad debt and credit memos*: revenue may be booked confidently while collections tell a more complicated story.
- *Cut-off issues*: revenue recognized a little too early, expenses recognized a little too late, especially around quarter-end when everyone wants the numbers to behave.

The point isn't to nitpick. It's to prevent a buyer from paying a multiple on earnings that were partially created by not recording the full cost of doing business. That's not "conservatism." That's a debt to the future, and future-you is the one who will pay it.

Revenue normalization: the most important question is ``why"

Normalizing revenue is less about math and more about causality. If revenue is up, is it because customers want more of the product, or because invoicing practices changed? If revenue is down, is it because demand fell, or because a one-time customer paused orders and will return?

The best diligence questions about revenue are the ones that force the seller to explain *mechanism*. "Revenue grew 18%" is a headline. Mechanism sounds like:

- Which customers expanded, and which contracted?
- How much growth came from price versus volume?
- Did growth come with margin, or did discounting do the heavy lifting?
- Are there customer concessions (free months, credits, extended payment terms) that aren't obvious in the top line?

When a seller says, "We had a one-time project," the follow-up isn't "prove it." The follow-up is: *What made it one-time?* If the company wins similar projects regularly, it may be recurring in spirit even if each project is individually non-recurring. Conversely, if the company benefited from an unusual market dislocation, a competitor's outage, or a customer's emergency spend, you may be staring at a revenue peak that looks stable only because you're standing on it.

EBITDA versus cash: the working capital trap

Even after you've agreed on Adjusted EBITDA, you can still misunderstand the business if you ignore working capital. Working capital is the operating cash tied up in day-to-day items like accounts receivable, inventory, prepaid expenses, and accounts payable. In deals, you'll hear "Net Working Capital" (NWC), which is simply current operating assets minus current operating liabilities (definitions vary, and that variation is the whole fight).

Here's the uncomfortable truth: two companies can have identical EBITDA and radically different cash behavior. One collects in 20 days and pays vendors in 45. Another collects in 75 days and pays vendors in 15. Same EBITDA, entirely different cash reality. If you buy the second one, you're not just buying earnings—you're buying a need to fund the business.

That's why the NWC "peg" (also called the target) becomes a negotiation hotspot. The peg is meant to ensure the buyer receives a "normal" level of working capital at closing—so the seller doesn't strip the business of cash by slowing vendor payments or accelerating collections in the final month. In spirit, it's simple: deliver the business with enough operating liquidity to run like it historically ran. In practice, it's a brawl over definitions and cut-offs.

The QoE helps by identifying unusual working capital patterns and by highlighting accrual issues that can distort NWC. But there's a key nuance: a QoE is built for insight, not for post-close true-up precision. The purchase agreement's working capital adjustment mechanism needs definitional clarity—what's included, what's excluded, which accounting policies apply, and how disputes get resolved. If diligence identifies an under-accrued bonus, that insight has to be converted into the agreement's accounting definitions, or it will show up later as a dispute where both sides claim the other is "changing the rules."

Debt-like items: the liabilities that don't introduce themselves

Another recurring feature of financial diligence is the search for "debt-like" items—obligations that behave like debt even if they don't sit neatly in a loan schedule. Buyers care because enterprise value is typically converted to equity value by subtracting net debt (again, definitions matter). If something is debt-like, it may reduce what the buyer is willing to pay for the equity.

Debt-like items often include:

- Unpaid taxes or payroll liabilities.
- Deferred revenue where cash has been collected but services haven't been delivered (especially important in software and services).
- Unfunded accruals for bonuses, commissions, and vacation.
- Litigation settlements payable, customer claims, or warranty obligations.
- Certain lease or equipment obligations, depending on deal definitions.

The reason these items cause so much friction is that they sit at the border between accounting categories and economic reality. Sellers may argue they're "normal operations." Buyers may argue they're "financing." Both can be partially right. The only way out is to decide, explicitly, how the purchase agreement will treat them—because otherwise the argument will simply be postponed until after closing, when it becomes more hostile and more expensive.

A QoE is also a management test-quietly, but unmistakably

One of the most useful byproducts of the QoE process is what it reveals about the finance function itself. Not every company needs a public-company finance team to be a good acquisition. But the buyer does need to know what they're inheriting: a disciplined close process or a fragile spreadsheet ecosystem held together by one person's heroics.

You learn this quickly by watching how the company responds to basic tie-outs:

- Can they reconcile revenue from the general ledger to customer-level detail?
- Can they explain why margins moved with something other than "mix"?
- Do their KPIs match their financial statements, or do they live in separate universes?
- When an error is found, do they correct it and document it, or do they argue it into exhaustion?

A seller that can answer clearly is not just making diligence smoother; they're reducing integration risk. A seller that can't reconcile fundamental numbers is effectively asking you to buy earnings on faith—and faith is not a line item lenders like to finance.

What ``good" looks like: defensible Adjusted EBITDA

By the end of financial diligence, the buyer should be able to say, with specificity, what Adjusted EBITDA means and why it deserves the multiple implied by the LOI. "Defensible" doesn't mean "perfect." It means the adjustments are explainable, supported, and not built on wishful thinking.

A practical way to sanity-check the outcome is to ask whether the adjusted number survives three uncomfortable questions:

- *If a skeptical lender reviewed this bridge, would they accept the logic?*
- *If the business had a bad quarter next year, would these adjustments still feel honest?*
- *If the founder vanished tomorrow, would the earnings power remain?*

If the answer is "yes" more often than "no," you're not just buying a number. You're buying a business with earnings you can actually operate.

A compact way to document adjustments: make the bridge auditable

Because add-backs can become a negotiation battlefield, it helps to present them in a format that forces clarity. A simple table can do what hours of debate cannot: make each adjustment declare what it is, why it exists, and what evidence supports it.

Adjustment	*Impact*	*Rationale (plain language)*	*Support (examples)*
Owner personal expenses	+0.9	Costs run through the business that do not relate to operations post-close	GL detail, invoices, policy, identified personal vendors
One-time legal fees	+0.6	Discrete matter not expected to recur	Engagement letter, invoice history, settlement documentation
Run-rate price increase	+0.4	Price change implemented, not fully reflected in trailing period	Customer notices, effective dates, billing extracts
Under-accrued vacation	-0.5	Earned employee benefit not fully recorded; likely cash outflow or liability	Payroll records, policy, accrual workpaper
Bad-debt reserve catch-up	-0.3	Historical collections suggest receivables are	AR aging, write-off history, collections notes

The value of this approach isn't the table itself. It's the habit: every adjustment earns its place. When adjustments are forced to show their work, the conversation becomes less about persuasion and more about economics.

The endgame: translating financial diligence into deal mechanics

Financial diligence doesn't end when the QoE report is delivered. It ends when the deal math is operational inside the purchase agreement and the closing model. That means:

- Adjusted EBITDA informs valuation, but it also influences earnout design if the parties go that route.
- NWC analysis becomes a peg/target and a post-close true-up mechanism with explicit definitions.
- Debt-like items become part of the net debt framework—again, defined, not implied.
- Identified liabilities become either purchase price adjustments, specific indemnities, closing deliverables, or post-close integration priorities.

If the diligence team can't translate its findings into those mechanics, the findings are just interesting trivia. And trivia is not what you pay bankers and accountants for.

When the translation is done well, something satisfying happens: the deal stops relying on trust and starts relying on design. The buyer isn't betting that everything will be fine; they're structuring the transaction so that if something is *not* fine, there's already a place for that reality to land—price, terms, or a plan. That's what a QoE is really for: not to make the seller look wrong, but to make the buyer less surprised.

3.3 Legal, Tax, and Compliance Reviews

A buyer can forgive a lot. Messy forecasting. A clunky CRM. Even margins that need work. What buyers don't forgive is discovering—after the wire hits—that the asset they bought came packaged with a lawsuit, a tax time bomb, or a contract that quietly gives the customer the right to walk. Legal, tax, and compliance diligence is where optimism goes to have its pockets turned out.

This workstream also has a peculiar emotional profile: it's not about making the business look good; it's about proving the bad things are either (a) not real, (b) bounded, or (c) priced and allocated. When it's done well, nobody celebrates. There's just a noticeable drop in everyone's blood pressure.

Corporate structure: does the company you're buying actually exist in the way the seller thinks it does?

Start with corporate housekeeping-not because it's glamorous, but because it's foundational. The fastest way for a deal to get stupid is to realize late that the company's legal structure doesn't match the economic story.

Buyers want to know: what entities exist, who owns them, and what those entities own. That sounds basic until you meet the real world. Subsidiaries get formed and forgotten. Old entities linger with open bank accounts. Equity gets issued casually in early years, documented with emails and hope. Options are promised in conversations and never papered. A "simple" cap table becomes a choose-your-own-adventure novel.

The goal here is not to shame anyone. The goal is to make sure the buyer is purchasing clean title to the equity (in a stock deal) or clean title to the assets (in an asset deal), without surprise co-owners showing up like uninvited guests.

Concrete documents matter: charter documents, bylaws or operating agreements, minutes/consents, equity grant paperwork, and a cap table that ties to reality. If there were prior financings, redemptions, or reorganizations, those aren't just historical trivia—they can affect who needs to sign, what approvals are required, and whether a "minor" stakeholder can create major friction at closing.

Authority and approvals: who has to say yes, and how many ways can they say no?

Every transaction has an invisible question behind it: *Can the seller actually do this deal?* Not "do they want to," but "do they have the legal authority and required approvals?"

That authority can be constrained in more places than people expect. Shareholder agreements may require supermajority votes. Investor rights documents may grant vetoes. Board approvals may be specific, procedural, and easy to botch. Debt agreements may prohibit a sale without lender consent. Even commercial contracts can function like a poison pill if they require counterparty consent and the counterparty knows you're desperate.

A good diligence team treats approvals like a gating item, not a footnote. Identify required consents early and put them on a timeline. Some consents are easy but slow (a landlord with a 30-day review cycle). Others are fast but unpredictable (a customer who uses consent as leverage for better pricing). The worst consents are the ones discovered late, when everyone is already mentally spending the purchase price.

Material contracts: reading for exits, not for poetry

Contracts are where a business's relationships become enforceable—and where they become fragile. Legal diligence on contracts is not an academic exercise in spotting defined terms. It's a hunt for escape hatches.

The highest-impact contract terms tend to fall into a few categories:

Change of control provisions

A change of control clause says, effectively, "If ownership changes, we get rights." Those rights vary. Sometimes it's a requirement to notify. Sometimes it's a right to terminate. Sometimes it triggers a pricing reset, accelerated payments, or the assignment restrictions that make a transaction structurally harder.

These clauses matter differently depending on deal structure. In a stock deal, the legal entity doesn't change; ownership does. In an asset deal, contracts often need assignment, which can trigger consent requirements even without explicit change of control language. Either way, the buyer wants to know: *Can we keep the revenue we're paying for?*

Most-favored-nation (MFN) and pricing provisions

MFNs sound commercially reasonable—customers like to know they're not being treated worse than their peers. In diligence, MFNs can be landmines. If the seller gave one customer a sweetheart deal, an MFN can quietly require similar treatment for others, compressing margins in a way the financial model didn't anticipate.

Service levels, credits, and remedies

If the business has performance obligations—uptime, delivery windows, defect rates—then the contract often contains credits or penalties when performance falls short. Buyers care because these terms turn operational hiccups into direct margin leakage. A company might have reported strong gross margin because it *rarely* issued credits, not because it *never* owed them.

Exclusivity, non-competes, and non-solicits

Exclusivity can be a moat or a shackle. Non-competes can protect the asset or invite litigation depending on enforceability and geography. Non-solicits can stabilize the workforce or be so narrow they're decorative. Each of these terms needs to be read with a practical eye: *What behavior does this contract actually allow after closing?*

One useful discipline: build a "top contracts" list early—top customers, top suppliers, key software and data providers, critical landlords—and require a structured abstract for each. Abstracting means pulling out the handful of terms that move the deal: term and renewal, termination rights, assignment/change of control, pricing, volume commitments, and dispute history. If you can't summarize a contract in a page, you probably don't understand it well enough to underwrite it.

Litigation and disputes: the difference between noise and a recurring pattern

Every business accumulates conflict the way ships accumulate barnacles. The question is whether disputes are occasional and containable, or whether they signal a deeper operating problem: product quality, employment practices, or a management team that defaults to fighting instead of fixing.

Legal diligence asks for schedules of claims, threatened claims, and settled matters. But the schedule isn't the whole story. The better question is pattern recognition:

- Are claims clustered around a product line?
- Do employment claims share a theme (overtime, discrimination, retaliation)?
- Does the company have repeat disputes with the same customers over performance or billing?

A single lawsuit can be a one-off. A pattern suggests you're buying a machine that produces liability, and the production line will keep running after you own it.

Buyers also care about insurance in a deeply practical way. Not "do they have insurance," but "does it respond?" Claims-made policies, exclusions, retroactive dates, and notice requirements can turn "covered" into "not covered" with depressing speed. If there's a known claim, diligence needs to know whether it was timely noticed and whether reserves reflect reality.

Intellectual property: do you own what you sell, and can you keep using what you use?

IP diligence is often described as "confirm ownership." That's accurate, but incomplete. The real risk is not just ownership; it's *continuity*. Can the buyer keep selling the product and operating the platform without paying surprise tolls—or being enjoined?

For software and technology-enabled businesses, the checklist is deceptively sharp:

- Are invention assignment agreements signed by employees and contractors?
- Were contractors used to build key code, and if so, does the company actually own that work product?
- Are open-source components used in ways that trigger licensing obligations inconsistent with the business model?
- Are trademarks registered, and are domain names actually held by the company (not the founder's personal account)?

In non-tech businesses, IP still matters—customer lists, proprietary processes, product designs, brand goodwill. The diligence point remains: you don't want to pay for something that can't be protected, or that requires ongoing permission from someone who didn't sign up to be your partner.

A small but common "gotcha" is ownership sitting in the wrong entity. The operating company sells the product, but the IP was assigned to a founder personally, or to an old entity that never got merged. Fixing that can be easy, but "easy" is not the same as "done," and buyers want it done before they pay.

Employment and benefits: liabilities that walk in every morning

People risks rarely announce themselves as "legal." They show up as turnover, morale issues, and operational fragility. But in diligence, the legal and financial sides intersect quickly: misclassification claims, unpaid overtime, benefit plan failures, and change-of-control obligations can all translate into real checks.

A few areas tend to carry disproportionate risk:

Independent contractor misclassification

Classifying workers as contractors instead of employees can reduce payroll taxes and benefits costs. It can also create liability for back wages, overtime, penalties, and taxes if the classification doesn't hold up. The diligence goal is not to debate philosophy; it's to assess exposure and likelihood. How were contractors used? Who directed their work? How long did they stay? Were they economically dependent on the company?

If a business's margins depend on aggressive classification, the buyer isn't just buying earnings—they're buying a compliance posture.

Change-of-control and retention obligations

Employment agreements, bonus plans, and equity plans can trigger payments at closing. Sometimes they're explicit ("transaction bonus"). Sometimes they're hidden in severance terms ("double trigger" or "single trigger" acceleration). These obligations can be material, and they can also affect integration: if key managers have an incentive to leave right after closing, your operating plan needs to account for that.

Benefits plan compliance

Retirement plans, health plans, and other benefits programs can carry compliance risk if testing, filings, or administration were mishandled. This is one of those areas where "we're small, we didn't know" is not a defense that regulators accept with enthusiasm.

Tax diligence: the silent partner that always gets paid

Tax diligence is often underestimated because the numbers look historical, the forms look bureaucratic, and the issues feel removed from operations. That complacency is expensive. Tax authorities have long memories, and tax exposures can survive closing with an indifference that feels personal.

Tax diligence asks two broad questions:

- *What's the exposure for prior periods?* Unpaid taxes, audits, nexus issues, penalties, unfiled returns.
- *How will the transaction change the future tax profile?* Structure choices, elections, and post-close compliance obligations.

The first question is about liability. The second is about deal design.

Historical compliance: "filed" is not the same as "clean"

A company can file returns and still have meaningful exposure. Sales tax is a frequent culprit because it's operationally messy and varies by jurisdiction. Payroll taxes can also create sharp risk because governments are particularly allergic to being unpaid.

A useful way to think about historical tax diligence is to look for mismatches:

- Revenue footprint versus where returns were filed.
- Workforce locations versus payroll tax registrations.
- Business model versus tax positions taken (e.g., services versus taxable products).

When mismatches appear, the buyer needs to decide whether the exposure is a rounding error, a manageable cleanup, or something that warrants escrow, indemnity, or a purchase price adjustment.

Structure-driven diligence: why "we can do a 338" is not a casual comment

Tax structure choices can turn into diligence requirements overnight. One example is a Section 338 election (including 338(h)(10) in certain contexts), which can cause a stock purchase to be treated more like an asset purchase for U.S. federal income tax purposes. The appeal is often step-up and amortization of assets—real economic value. The cost is complexity and a need for precision.

When a structure election is on the table, tax diligence becomes intensely detail-oriented about ownership and timing. Rules around what qualifies (for example, whether a buyer has purchased enough stock within the relevant period) depend on cap table accuracy, stock movements, redemptions, and the acquisition timeline. That's not just a tax technicality—it's a diligence mandate. If the cap table is "mostly right," tax planning can become "mostly wrong."

There's also a practical, human element: elections may require specific parties to file and sign, sometimes across corporate groups or shareholder bases. If signatures are hard to gather—or if someone has leverage and knows it—that can affect timeline and deal certainty. Tax diligence is where those constraints are surfaced early enough to manage, rather than discovered when everyone is already aiming for a signing date.

Compliance diligence: successor liability and the problems you inherit by owning the asset

Compliance diligence can feel like a niche concern until you remember a basic truth: regulators don't care that you're the "new owner." Depending on the area—anti-corruption, sanctions, environmental, data privacy, industry-specific regulation—liability can attach to the company, the assets, or the conduct. Ownership changes, but responsibility often sticks.

The compliance lens is particularly sharp in cross-border business, government-facing revenue, healthcare, financial services, and any business with meaningful third-party agents or distributors. If revenue depends on intermediaries, gifts, travel, rebates, or "consulting" arrangements, compliance diligence becomes less about policies and more about behavior.

A strong compliance review doesn't just ask, "Do you have a policy?" It asks:

- Who sells, and how are they compensated?
- Who are the third parties, and what do they actually do?
- Where are the pressure points—countries, customers, licensing processes—where shortcuts are tempting?
- Have there been investigations, hotline complaints, or internal incidents, and how were they handled?

A subtle but critical point: even when pre-close diligence is constrained by time or access, compliance risk doesn't disappear. It shifts into a post-close obligation. That affects integration planning and may influence what protections the buyer asks for in the definitive documents. A buyer that treats compliance diligence as "check the box" is effectively choosing to learn about the company's risk posture after they own it, when the learning is more expensive.

Cyber and data: the liability that doesn't show up as a lawsuit until it does

Cyber diligence used to be considered a technical specialty. Now it's basic corporate hygiene. A serious security incident can create direct costs (response, remediation, ransom demands) and indirect costs that are often worse (customer churn, regulatory inquiries, contractual penalties, reputational damage).

The legal/compliance angle is not "can the IT team patch servers." It's:

- What data is collected and stored (PII, payment data, health data)?
- What obligations exist under customer contracts regarding security and breach notification?
- What has happened historically—incidents, near-misses, insurance claims?
- Are there policies, training, and controls that a reasonable buyer can stand behind?

If the company is in a regulated space, the bar is higher. If the company sells to enterprise customers, the contractual bar is higher. In either case, the diligence output isn't just an IT report—it's an assessment of whether the buyer is walking into a situation where one breach can turn into multiple lawsuits and a very bad quarter.

How legal, tax, and compliance findings become deal terms

The most frustrating diligence outcome is discovering a real issue and then watching it evaporate into vague language. Issues need to be translated, and legal/tax/compliance issues have a fairly standard menu of translations:

- *Condition to close*: fix the problem before closing (e.g., obtain a critical consent, clean up IP assignments).
- *Purchase price adjustment*: reduce price for quantified exposure.
- *Specific indemnity*: allocate a known risk to the seller with tailored survival periods and caps.
- *Escrow/holdback*: create a funded backstop for known or suspected liabilities.
- *Covenant*: require certain actions pre-close or post-close (e.g., compliance remediation, filing delinquent returns).

The "right" translation depends on the nature of the risk. A missing IP assignment is often best handled as a condition to close: don't buy until you own what you're buying. A sales tax exposure might be handled through indemnity and escrow if it's quantifiable but not fully certain. A change-of-control consent might be a condition to close if the customer is essential, or it might be a walk-away trigger if the customer refuses and the revenue concentration is unacceptable.

Good diligence teams keep a running list of these translations alongside the findings. That prevents the classic deal drift where everyone agrees something is a problem, but nobody agrees what to *do* about it, and then the timeline forces a decision made by exhaustion rather than judgment.

A final reality check: the best ``legal diligence" often sounds like operational common sense

It's tempting to treat legal, tax, and compliance diligence as a separate universe run by specialists. But the best findings are often plain-English statements about how the business actually behaves:

- "Two customers can exit on 30 days' notice if ownership changes."
- "The company's best engineer wrote key code as a contractor, and ownership is unclear."

- "Sales tax was never filed in states where there are clear sales footprints."
- "Third-party agents are paid in ways that create compliance risk, and oversight is thin."

Those are not legal trivia. They are business facts with legal consequences—and financial consequences right behind them. When diligence produces that kind of clarity, the buyer isn't just reducing downside. They're buying with their eyes open, which is the only respectable way to buy anything expensive.

3.4 From Findings to Schedules: Managing Disclosure

Diligence produces facts. Agreements allocate consequences. Disclosure schedules are the narrow bridge between the two—and theyŕe easy to underestimate because they look like paperwork. In practice, schedules are where a seller turns "yes, that ugly thing exists" into "and you canť act surprised later." They are defensive writing with real teeth: a list of exceptions to the sellerś representations and warranties (the promises in the purchase agreement about whatś true).

The emotional tone shifts here. Earlier, the buyer is asking questions like a detective. Now the seller is answering like a lawyer. Not because the seller suddenly became uncooperative, but because the rules changed: once an issue is properly disclosed, it often stops being a breach. The seller may still have to fix it, pay for it, or negotiate around it—but disclosure changes the leverage. It transforms a hidden defect into an acknowledged condition of the deal.

What disclosure schedules actually do (in plain English)

Representations and warranties ("reps") are the sellerś assertions: we own the assets, contracts are enforceable, financial statements are not misleading, thereś no undisclosed litigation, employees were paid properly, taxes were filed, and so on. The buyer relies on those promises in deciding to close. If a rep is false, the buyer may have a claim after closing—subject to caps, baskets, survival periods, and the rest of the risk-allocation machinery.

Disclosure schedules qualify those promises. They are attachments to the agreement where the seller lists exceptions: lawsuits that exist, contracts with change-of-control clauses, liens on assets, related-party transactions, customer disputes, open-source licenses, tax audits, benefit plan issues—whatever makes a rep technically or practically untrue.

The critical mechanic is this: a disclosure usually turns a "breach" into a "known issue." Known issues dont́ magically become harmless, but they often stop being actionable as a rep-and-warranty claim. Said differently, if you disclose that the top customer can terminate upon a change of control, you may have just prevented the buyer from suing later on the theory that the contract was freely transferable. The buyer can still decide the issue is unacceptable, demand a consent, demand a price adjustment, or walk. But the seller is no longer "hiding the ball."

Thatś why schedules are not clerical work. They are a negotiation over what risks are *accepted* versus what risks remain *warranted*.

The sellerś incentive is not to be honest---itś to be immunized

A seller can be sincere and still have a strategy. The strategy is not "disclose everything," and itś not "disclose nothing." Itś "disclose enough to protect ourselves without scaring the buyer or giving away leverage."

That creates a predictable tension. Sellers prefer broad, general disclosures because breadth feels safe: one line that says "There may be disputes with customers from time to time" sounds like it covers a lot. Buyers prefer specific, targeted disclosures because specificity is enforceable. "Customer X sent a notice of breach on January 12, 2026 alleging missed service levels and seeking $480,000 in credits" is the kind of sentence that changes negotiation dynamics—precisely because itś concrete.

Schedules are where these preferences collide, and the winner is usually the party with the better process and more stamina.

A disclosure schedule is only as good as the rep it qualifies

This is the first uncomfortable truth about schedules: they dont́ live on their own. They only matter in relation to the reps and the agreementś "rules of interpretation" for disclosures.

If a rep says, "No material customers have the right to terminate upon a change of control," then the schedule must disclose those customers and those rights in a way that actually qualifies that rep. If the schedule discloses something related but not quite on point—say, it lists "Material Contracts" without stating the termination right—the seller might not get the protection it thought it bought.

Conversely, buyers can get lulled into complacency by volume. A schedule can be hundreds of pages. It can feel like disclosure is happening because the document is heavy. But weight is not precision. The buyer needs to connect each meaningful diligence finding to the specific rep it would otherwise contradict, and then decide whether the rep should be carved out, tightened, or left alone with a requirement to fix the issue.

Cross-disclosure: the magic trick that sometimes works and sometimes doesnt́

Sellers love cross-disclosure. Cross-disclosure is the idea that if a fact is disclosed somewhere in the schedules, it should qualify multiple reps—even if it wasnt́ repeated under each relevant schedule section. The logic is efficiency: why list the same lawsuit in five different places?

The problem is that contracts arenť sympathetic to efficiency unless the drafting makes them sympathetic. Whether cross-disclosure works depends on the agreementś language and the quality of the disclosure. Many agreements contain "global" language along the lines of: a disclosure in one schedule qualifies other reps if the relevance is reasonably apparent on its face. That phrase—"reasonably apparent"—does a lot of work, and it does it ambiguously.

Ambiguity is where litigation thrives. Buyers and sellers can look at the same fact and disagree about whether it was "reasonably apparent" that it qualified a different rep. If the seller disclosed a list of contracts and one of them contains a change-of-control clause, was it reasonably apparent that this qualified the rep about change-of-control termination rights? Maybe. But if the contract list doesnť flag the clause and the clause is buried on page 47, a buyer will argue it was not reasonably apparent. A seller will argue the buyer had the contract and could read it. Both arguments can be made with a straight face. Thatś the issue.

The practical best practice—especially for high-stakes items—is not to rely on cross-disclosure. Repeat the disclosure in the specific schedule section that maps to the rep, or include an explicit cross-reference that leaves no room for interpretive gymnastics. Redundancy is cheap compared to a post-close dispute.

The word ``none" can be a landmine

Schedules often have sections with headings like "Litigation," "Environmental Matters," "Material Contracts," "Compliance with Laws," and the seller fills in entries—or writes "None."

The seller may mean "none that are material." Or "none that we know about." Or "none other than what we disclosed elsewhere." But the schedule entry says "None," and contracts tend to read words literally when money is at stake.

If the agreement doesńt clearly state that disclosures in one place qualify other reps, a "none" entry can create a trap: it can be used to argue that the seller affirmatively stated there were no exceptions under that rep, even if somewhere else the schedules mention a related fact. Even worse, inconsistent practice across schedules—sometimes cross-referencing, sometimes saying "none," sometimes being silent—can create an impression that the seller was being selective rather than systematic.

The fix is not complicated, just disciplined:

- Avoid "none" when "none" isńt literally true.
- Use consistent cross-referencing conventions if cross-disclosure is intended.
- Where something matters, disclose it w.here it belongs, not where itś convenient.

This is one of those unglamorous drafting points that can decide whether a buyer has a claim later. One word can make a schedule either a shield or a boomerang.

Turning a diligence finding into a schedule entry: the ``bad fact'' needs a clean sentence

A diligence finding often starts life as a messy bundle: emails, explanations, half-remembered history, and a couple of documents that dont́ quite match. The schedule entry has to compress that mess into something legally usable.

A good schedule entry usually contains:

- The identity of the thing (party name, contract name, case caption, regulator).
- The nature of the exception (what rep would be untrue without this disclosure).

- Enough detail to be meaningful (termination rights, dollar amounts, dates, allegations).
- Attachments or pointers (the actual contract, notice letter, or correspondence).

A bad schedule entry hides behind vagueness: "Customer dispute exists." That's not a disclosure; it's a weather report. A buyer can't price it, can't model it, and can't decide whether it's a deal-breaker. And if the seller's goal is immunity, vagueness can backfire—because a court may decide the disclosure wasn't adequate to qualify the rep.

There's a strange discipline to writing an effective "bad fact" sentence. It should not be editorial. It should not minimize. It should be factual, specific, and calm. The tone you want is: *We are not embarrassed, we are not hiding, we are disclosing*. That tone is protective.

Schedules as negotiation: disclosure doesn't end the conversation---it relocates it

A seller sometimes acts as if disclosure is the end of the story: "We disclosed it, so it's fine." A buyer sometimes acts as if disclosure is a confession: "You disclosed it, so you're guilty." Both are wrong.

Disclosure changes what happens next. It doesn't eliminate negotiation; it gives negotiation a more accurate map.

Once a fact is disclosed, the buyer has a menu of responses:

- *Accept it* (and model it as part of the business).
- *Fix it pre-close* (get the consent, clean up the IP assignment, terminate the problematic vendor).
- *Price it* (purchase price reduction, working capital target adjustment, debt-like treatment).
- *Isolate it* (specific indemnity, escrow/holdback).
- *Re-trade structure* (asset deal instead of stock deal; carve-outs; exclusions).

The seller also has choices. A seller can volunteer to cure issues to preserve price. A seller can resist and gamble that the buyer wont́ walk. A seller can offer narrow protections and insist the rest is part of buying a real company. The schedules are the chessboard where those choices become visible.

The interface with financial diligence: definitions matter more than accusations

Some of the most important "schedule-like" disclosures arent́ even in the schedules. They live inside definitions: what counts as "Indebtedness," what counts as "Net Working Capital," whatś excluded from working capital, whatś treated as transaction expenses, and how accounting policies are applied in the closing statement.

This is where financial diligence findings must be converted into legal mechanics. If diligence found under-accrued bonuses, the buyer can either:

- Fight about it as a "breach" (often harder, because the seller will disclose it).
- Or treat it as a working capital or debt-like adjustment (often cleaner, because it becomes math).

Math is not always friendly, but it is at least structured. A purchase agreement can define, precisely, that accrued bonuses are included in NWC (or in debt-like items), measured consistently, and subject to a true-up. That converts a squishy debate into a post-close mechanism.

This is why strong deal teams dont́ view schedules as isolated legal attachments. They view the entire agreement—definitions, covenants, closing mechanics, schedules—as one integrated system for allocating what diligence uncovered.

How buyers get burned: the ``we had it in the data room" defense

Sellers sometimes argue that uploading something into the VDR should count as disclosure. Buyers sometimes assume that if something was uploaded, it must have been disclosed. Both assumptions are dangerous.

A VDR is a warehouse. A schedule is a label. Warehouses can contain dangerous chemicals; the label is what tells you theyŕe there.

Whether "data room disclosure" qualifies reps depends on the agreement. Some deals explicitly say that only the schedules disclose exceptions. Others say that documents made available in the data room can qualify reps, sometimes subject to conditions (for example, the documents must be indexed and clearly identified). If the agreement is silent or ambiguous, youv́e created an interpretive fight.

The buyer-friendly approach is simple: if it matters, it should be on the schedules, even if the supporting document is already in the VDR. The seller-friendly approach is also simple: if you want the data room to qualify reps, negotiate clear language to that effect and maintain a clean index. Either way, vagueness is not your friend.

Schedule hygiene: consistency is a form of credibility

Schedules are usually drafted under time pressure, which is exactly when sloppy practices multiply. But schedules are also a credibility signal. A buyer can feel when a seller is disclosing systematically versus dumping documents and hoping for the best.

A few habits create schedule hygiene:

- *Consistent naming*: parties, contracts, and entities are labeled the same way throughout.
- *Consistent thresholds*: if "Material Contract" has a definition, the schedule shouldnt́ quietly use a different standard.

- *Consistent cross-referencing*: either use it carefully or don't use it; inconsistency invites adverse inference.
- *Document completeness*: if a contract is scheduled, the final executed version and all amendments should be in the VDR and easy to locate.

This isn't aesthetic. It affects enforceability. If schedules read like a coherent database, they're more likely to be treated as intentional disclosures. If they read like a collage, they're more likely to be treated as inadequate—especially when a buyer claims it was misled.

The buyer's playbook: run schedules like a mapping exercise

The buyer's job is not to read schedules like literature. It's to map them.

A practical method is to create a "rep-to-disclosure map" that lists each rep and the disclosed exceptions that qualify it. This does two things. First, it prevents surprises: you can see which reps are being carved out heavily and which are still largely clean. Second, it forces prioritization: not every disclosure matters equally, and the map helps you focus on the ones that impact valuation, integration, or legal exposure.

This is also where the earlier diligence triage shows its value without being re-litigated. Red-flag findings should trigger schedule entries that are unmistakable, paired with deal terms that allocate the risk intentionally. Yellow-flag findings often belong in schedules and in integration plans, but they should not be allowed to quietly hollow out the reps without compensation. A rep that is carved out so thoroughly that it becomes meaningless is not "standard." It's a negotiated concession, and it should be treated like one.

The sellerś playbook: disclose precisely, and dont create new problems

Sellers sometimes over-disclose in a way that creates confusion. A schedule that includes every minor complaint can make the company look litigious. A schedule that attaches draft documents can create disputes about what was actually agreed. A schedule that references outdated contracts can accidentally disclose something irrelevant while failing to disclose the real issue.

Precision is the sellerś friend. Disclose the real exceptions, with enough detail to qualify the rep, without turning the schedules into a landfill. And where the seller wants broad protection, the right way to seek it is not vague schedule entries—itś negotiating the reps themselves: materiality qualifiers, knowledge qualifiers, and appropriate scopes. Trying to solve drafting problems with sloppy schedules is like trying to solve plumbing problems with paint.

Why this matters after closing: schedules become the battlefield record

After closing, disputes tend to have a predictable script. Something bad happens. The buyer feels it was misled. The seller insists it was disclosed. Then everyone starts reading the purchase agreement like itś scripture, hunting for the sentence that decides who pays.

In that moment, disclosure schedules stop being attachments and become evidence. They are the record of what the seller admitted, what the buyer accepted, and what both sides chose to leave as a warranted promise. A well-crafted schedule reduces ambiguity and lowers the odds that a post-close disagreement becomes a claim. A sloppy schedule increases ambiguity and practically invites a fight—because each side can plausibly read it in their favor.

The quiet genius of good disclosure management is that it makes the deal sturdier without making it louder. It replaces "trust us" with "here are the facts," and then forces those facts to be priced, fixed, or contractually allocated. Thatś not just legal hygiene. Thatś deal mechanics doing their job.

Chapter 4

The Definitive Agreement: Economics and Risk

If the Letter of Intent was the engagement ring, the Definitive Agreement is the marriage license-and the pre-nup. This isn't just a formality for lawyers to bill hours against; it is the comprehensive rulebook that translates business promises into enforceable mechanics. In this chapter, we strip away the dense legalese to reveal the document's true purpose: defining exactly what is being bought, how much is actually paid (which often differs from the headline price), and who pays the bill when things break after closing.

4.1 Inside the Purchase Agreement: The Chassis and the Accessories

The definitive agreement is where a deal stops being an idea and becomes a machine. Until this point, everyone has been speaking in headlines: valuation, structure, timing, "we're aligned." The purchase agreement is what happens when you force those headlines to survive contact with reality. It takes the squishy parts of the bargain—what exactly is being sold, what comes with it, what doesn't, what happens if something is wrong, what happens if someone stalls—and bolts them down with definitions, procedures, and consequences. If the LOI is a handshake with a price tag, the definitive agreement is the torque spec.

The easiest way to understand the document is to treat it like a chassis: the rigid frame that everything else attaches to. A chassis isn't the part you admire in a showroom, but it determines whether the car drives straight, rattles, or folds in a crash. In M&A, that "frame" is usually one of three formats: a Stock Purchase Agreement (SPA), an Asset Purchase Agreement (APA), or a Merger Agreement. The names change, and the tax/legal mechanics shift, but the function is the same: specify *what* is transferring, *how* it transfers, *what* the buyer is paying, *when* money and ownership move, and *who eats the loss* if the picture you were sold turns out to have a crack.

You can often tell the real economic story of a transaction by reading the table of contents before you read a single sentence. Purchase agreements are not written like novels; they are written like operating manuals, with the important parts hidden in plain sight. The structure tends to look deceptively standardized: definitions; the transaction steps; consideration; representations and warranties; covenants; conditions to closing; termination; miscellaneous boilerplate. The twist is that the "standard" outline is a clever disguise for a document that can behave very differently depending on a handful of loaded choices. Two agreements can have identical headings and radically different outcomes because the definitions under those headings do all the work.

Start with the deceptively humble opening act: *Definitions*. In normal writing, definitions are where you park boring clarifications. In M&A, definitions are where you hide explosives. "Indebtedness" isn't a dictionary word; it's a negotiated universe. Does it include capital leases? Letters of credit? Earnout obligations? Change-of-control bonuses? Unpaid transaction expenses? The buyer's check at closing is often computed by taking some headline number and subtracting "Net Debt" and "Transaction Expenses" as those terms are *defined*. If you want to understand why sophisticated people fight over a few lines of text at 2:00 a.m., picture this: a single defined term can swing millions of dollars without changing the "purchase price" the parties boast about in the press release (or the email they send to their teams).

Those definitions also control what the parties are actually promising to deliver. "Material Adverse Effect" (or "MAE") is another phrase that looks like common sense until you try to use it. The document will typically define it, then define what *doesn't* count (general economic changes, industry downturns, acts of war, pandemics, interest rates), sometimes with a final catch: unless the target is hit *disproportionately*. That last word can turn a market crash into a company-specific problem—or vice versa—depending on how it's drafted. Even when everyone believes they will never litigate, the agreement is quietly preparing for the day someone needs a lever big enough to move the world.

After definitions, the document usually moves to the part that feels like the "deal" to a non-lawyer: *The Transaction Mechanics*. This is where the agreement names the parties, identifies the structure, and spells out the steps. In a stock deal, the buyer purchases equity from the sellers; the legal entity continues, still owning its contracts, IP, permits, and employees. In an asset deal, the buyer purchases specified assets and assumes specified liabilities, often leaving behind unwanted obligations—at least in theory. A merger agreement can resemble a stock deal economically, but the legal path is different: one entity merges into another, sometimes used to cleanly squeeze out minority holders, simplify approvals, or fit a particular corporate strategy. The chassis you choose determines how many parts have to be detached and reattached at closing.

That choice has a personality. Stock deals are operationally clean: the company remains itself, and most contracts stay in place without assignment language—though change-of-control clauses can still trigger consent requirements. Asset deals can be wonderfully precise and brutally tedious. Precise because the buyer can try to cherry-pick: these assets, those contracts, this inventory; not that litigation, not that old tax year, not those warranties you issued to customers five years ago. Tedious because you have to list and transfer everything you want, and sometimes you don't know what you forgot until you discover it was holding up a revenue stream. Asset deals also carry an uncomfortable truth that the contract can't entirely wish away: certain liabilities can follow assets as a matter of law, even if the agreement says "excluded." A well-drafted APA manages that tension; it does not pretend it doesn't exist.

The mechanics section also reveals whether the deal is *sign-and-close* or *sign-then-close*. A sign-and-close transaction is like buying a house in cash with the keys handed over the same day: once the ink dries, ownership and money move almost immediately. A sign-then-close deal creates a time gap between signing and closing. That gap can be a few days or many months, and the purchase agreement becomes a bridge over that gap. The longer the bridge, the more engineering you need: covenants to run the business "in the ordinary course," closing conditions, regulatory approvals, consents, financing commitments, and a detailed map of what happens if something breaks.

The document's third major area is *Consideration*: what the buyer is paying, to whom, in what form, and under what math. Even when the headline number looks fixed, the agreement usually breaks it into components: cash at closing, rollover equity, seller notes, escrows, holdbacks, and sometimes contingent payments. The legal drafting here often feels weirdly ceremonial—the agreement might list the "Purchase Price" as a single number and then spend pages explaining why that number is not what will be wired. This is not a drafting failure; it's a reflection of the fact that "price" in M&A is a mechanism, not a sticker.

Consideration also introduces the closing flow of funds, which is where the agreement quietly turns into choreography. Who pays transaction expenses? Do they come out of the purchase price or get paid separately? Are there third-party payoffs (banks, lessors, tax authorities) that must be satisfied at closing? Is there an escrow agent that needs wiring instructions, an escrow agreement, and rules about release? If you ever want to see the difference between a theoretical deal and a real one, watch how long experienced counsel spends on the closing payment mechanics. Money is the most honest witness in the room.

Then comes the part of the agreement that behaves like a security system: *Representations and Warranties*. Think of these as the seller's signed, contract-grade snapshot of the business: financial statements are accurate in a specified way, taxes have been paid, key contracts are valid, there's no undisclosed litigation above a threshold, the company owns its IP, the company complies with laws, and so on. The point is not that the buyer doesn't know anything; the point is that the buyer wants the seller to *stand behind* specific facts. When a rep turns out to be wrong, the buyer's remedy is usually found elsewhere in the document (indemnification) or in an insurance policy (RWI), but the rep itself is the trigger.

Reps are also where the agreement tells you what kind of diligence actually happened. Not through a confession, but through the seller's schedules and carve-outs. The schedule disclosure is the seller's way of saying: "Yes, we promised X, but here are the exceptions." Those exceptions can be innocent (a list of minor lawsuits), alarming (a regulatory inquiry), or purely definitional (what exactly counts as "Material Contract"). A buyer who reads schedules like an archaeologist reads soil layers will learn what the business is most worried about—because those are the things the seller insisted on disclosing with surgical precision.

After reps come *Covenants*, which are promises about conduct. Some covenants are pre-closing: run the business in the ordinary course; don't fire key people without consent; don't issue equity; don't take on debt; keep insurance in place; preserve relationships with customers; keep books consistent. Some covenants are post-closing: non-competition, non-solicitation, confidentiality, cooperation with tax filings, access to records, transitional support. Covenants are where deals become psychologically hard, because they force the seller to live with constraints *after* mentally moving on. They force the buyer to tolerate a company it doesn't yet own behaving like a company it already owns. The agreement manages that tension by turning it into a list of permissions and prohibitions. It's not romantic, but it prevents the kind of drift that turns a signed deal into a lawsuit.

A particularly important covenant category is *Access and Cooperation*. Between signing and closing, the buyer may need help obtaining consents, pushing regulatory filings, getting landlords to sign, persuading customers not to panic, and planning integration. Sellers can be tempted to treat that period as a waiting room. Buyers can be tempted to start rearranging furniture before they own the house. Good drafting draws a clear boundary: cooperation yes, control no. That boundary also matters for risk: if the buyer is deemed to be controlling the business before closing, liabilities can follow in uncomfortable ways.

The agreement then gets to what feels like the "escape hatches": *Conditions to Closing* and *Termination*. Conditions are not merely legal boxes to tick; they are leverage. A closing condition might be that the seller's reps are true at closing, that no MAE has occurred, that required consents are obtained, that key employees have signed agreements, that financing is available, that a government authority has cleared the transaction. Each condition is a sentence that can, in the wrong moment, become a crisis. The art is to draft conditions tight enough to protect against real disasters and loose enough that ordinary messiness doesn't hand one party a weapon.

Termination provisions are the agreement's breakup protocol. They set a drop-dead date (outside date), explain what happens if closing hasn't occurred by then, and determine whether anyone owes a termination fee or expense reimbursement. In private deals, termination fees are less common than in public deals, but not unheard of, especially where the seller wants to prevent the buyer from treating the signed agreement as a free option. The termination section is also where you'll find provisions about specific performance—the legal power to force someone to close, not just pay damages. That single concept can turn a "good faith effort" into a "you are going to closing unless the sky falls."

All of this still describes the chassis. But deals don't run on a chassis alone. They run because a series of ancillary agreements—the accessories—attach the economic bargain to specific people, assets, and practical realities. These documents often look "secondary" because they're shorter and sometimes negotiated later in the process, but they are frequently where the deal becomes operationally true. If the purchase agreement is the frame, ancillary agreements are the tires (traction), the engine (incentives), and the wiring harness (transitional dependencies).

One accessory shows up in almost every private-company deal with real employees: *Employment and Incentive Arrangements*. Buyers rarely buy companies; they buy the ability of people to keep producing results the morning after closing. That ability is fragile. A founder who has been sprinting through diligence can become strangely calm once the wire hits, and a key sales leader can decide that their newfound LinkedIn value is irresistible. Employment agreements, offer letters, retention bonuses, and equity rollovers are not just HR niceties; they are continuity devices. They create a post-closing narrative for the people who hold the customer relationships, product roadmap, and institutional memory in their heads.

These arrangements also sneak in the most sensitive negotiation in the whole deal: *control*. Will management report to the buyer's CEO, a business unit head, or stay semi-autonomous? Are there performance metrics? Is there a severance package if integration turns into replacement? The purchase agreement may say little about this, but the side letters say everything. A seller may accept a lower price for the dignity of a clear role and a stable team. A buyer may pay more for a management structure that doesn't implode in month two.

Then there are *Restrictive Covenants*: non-compete, non-solicit, non-disparagement, confidentiality. These are the accessories that keep the seller from selling the same business twice—first to the buyer, then again by recreating it across the street. The enforceability of these covenants depends heavily on jurisdiction, scope, and consideration, and the drafting has to respect that reality. In many deals, restrictive covenants are split across multiple documents: some in the purchase agreement (binding the sellers), some in employment agreements (binding key employees), and sometimes a separate restrictive covenant agreement for a founder who is rolling equity but no longer operationally employed.

Non-competes deserve a special kind of respect because they are simultaneously central and controversial. Central because a buyer often pays for customer relationships and know-how that can walk out the door. Controversial because the legal system doesn't love restraints on trade, and regulators have taken an increasingly skeptical view in some contexts. Smart deal teams treat restrictive covenants like a precision instrument: tailored to protect legitimate business interests rather than a broad attempt to silence competition. A covenant that is too aggressive can fail when you need it most; a covenant that is too timid can turn the buyer's goodwill into the seller's seed funding.

Another accessory is the *Escrow Agreement*. In many private deals, a portion of the purchase price is held back in escrow to secure post-closing obligations: indemnification claims, purchase price adjustments, sometimes tax matters. The escrow agreement is the rulebook for the escrow agent: who can give release instructions, what evidence is required, how disputes are handled, and when funds are automatically released. This document can feel like plumbing, right up until you have a dispute. Then it becomes the only thing that matters, because escrow agents are professional neutral parties whose favorite word is "no." They do not interpret your intent; they follow the release mechanics. If the release mechanics are vague, everyone learns what "vague" costs.

Where escrows exist, a *Stockholder Representative* (or seller representative) agreement often follows. In a dispersed cap table—many individual sellers, options holders, maybe a fund exiting—the buyer needs one counterparty to receive notices, negotiate claims, and sign off on adjustments. The representative is essentially a deal-time role that lives post-close, sometimes for years, with the authority to bind the selling group. This can be a thankless job: the representative gets pressure from the buyer (pay the claim) and from the sellers (fight the claim) and is often making decisions with imperfect information. The purchase agreement will define that authority; a side agreement may address compensation, indemnification of the representative, and the internal rules among sellers. When this is drafted well, the buyer has one door to knock on. When it's drafted badly, the buyer knocks on fifty doors and gets sued for knocking.

Transition Services Agreements (TSAs) are another classic accessory, especially in carve-outs. If a business is being separated from a larger company, it may not have its own standalone IT systems, payroll, benefits, accounting, or even customer billing. A TSA is the temporary scaffolding that keeps the business from collapsing while the buyer builds (or plugs in) permanent infrastructure. TSAs can also appear in non-carve-out deals when the seller has capabilities the target relied on informally. The TSA answers unglamorous questions that determine whether the acquired company can invoice customers next week: Who hosts email? Who administers health benefits? Who holds the customer data? What service levels apply? How much does the buyer pay, and when does the service end?

TSAs are also where you see the shadow of leverage. Sellers don't want to be trapped providing services indefinitely; buyers don't want a cliff. The agreement might have extension options, but the price for extensions is usually designed to feel unpleasant. That unpleasantness is intentional: it creates urgency. The TSA is not there to create a new business relationship. It's there to buy time without buying excuses.

IP Assignment and *Invention Assignment* agreements are accessories that matter disproportionately in tech and product-heavy businesses. A buyer may believe it is purchasing the company's crown jewels—code, trademarks, patents, trade secrets—only to discover that the chain of title is cloudy. Perhaps a contractor built part of the product and never assigned rights. Perhaps a former employee's agreement is missing. Perhaps open-source software was used in a way that creates licensing obligations. The purchase agreement will contain reps about ownership and non-infringement, but the ancillary documents often do the cleanup: specific assignments, confirmatory transfers, moral rights waivers where relevant, and new invention assignment agreements with continuing employees. In other words: the chassis promises the engine exists; the accessory hands you the keys.

Real estate and *contract assignment* documents are the accessories that make an asset deal feel like moving house—in a storm—while the house is still occupied. Leases may require landlord consent; customers may require consent; software licenses may be non-transferable; permits may not move cleanly. The purchase agreement can state that the seller will use "reasonable best efforts" to obtain consents, but the actual transfer happens through separate instruments: assignment and assumption agreements, landlord consents, estoppels, novations, sublicenses, and sometimes new contracts signed at closing. These accessories are where a deal quietly risks delay. If a key customer consent is missing, is closing conditioned on it? If not, what happens if the customer walks post-close? If it is conditioned, who bears the risk of never obtaining it? The answer is not philosophical; it's a schedule and a closing condition.

Debt payoff letters and *lien releases* are accessories that matter because they convert "we'll pay the debt" into legal finality. In a cash-free/debt-free deal construct, the buyer expects to receive the business without financial encumbrances. That doesn't happen because the agreement says so; it happens because lenders deliver payoff letters, UCC termination statements are filed, mortgages are released, guarantees are dealt with, and escrow or closing counsel routes funds to the right places. A deal can be economically perfect and still fail if lien releases are not buttoned up. The buyer isn't paying to inherit a bank's security interest.

Equity rollover documentation is an accessory that often carries more emotion than the purchase agreement itself. When a founder or management "rolls" equity, they are converting part of their sale proceeds into a continuing stake. That can align incentives beautifully—or create years of friction if governance and economics are unclear. The rollover documents can include subscription agreements, operating agreements, shareholder agreements, drag-along rights, vesting terms, repurchase rights, and information rights. The purchase agreement may say "Seller will roll 20%" as if that's a simple sentence. The rollover documents reveal what 20% *means*: minority protection or minority exposure; a voice or a silent seat; a path to liquidity or a locked box.

The accessories also include the *Disclosure Schedules* themselves, which are sometimes treated like an appendix rather than a document. That's a mistake. Schedules are where the seller tells the truth in a way that is admissible, organized, and legally meaningful. They list contracts, employees, benefit plans, litigation, IP, customers, compliance matters, related-party transactions, and exceptions to the reps. They can run hundreds of pages. In the worst deals, schedules are assembled like a school project the night before it's due. In the best deals, schedules are treated as a final diligence exercise: comprehensive, consistent with the reps, and aligned with how the business actually operates. When a post-close dispute arises, the schedules are often the first place everyone looks—not because they're dramatic, but because they are precise.

A purchase agreement also creates a kind of ecosystem at signing and closing: officers' certificates, bring-down certificates, legal opinions (sometimes), FIRPTA certificates in certain contexts, secretary certificates, good standing certificates, and a closing memorandum that functions like an air-traffic-control plan for signatures and wires.

These are accessories you don't notice when they work. When they fail, closing feels like trying to land planes in fog.

One of the most underappreciated features of the chassis is that it doesn't just say what happens at closing; it creates *post-closing behavior*. It appoints who can speak for the sellers, sets deadlines for post-closing adjustments, defines notice procedures, allocates tax responsibilities, and sometimes dictates the forum and process for disputes. People think of M&A as a moment. The definitive agreement treats it as a relationship with a breakup clause already installed, just in case. That may sound cynical, but it's actually optimistic: the document assumes that life will continue, surprises will occur, and the parties would like a way to handle those surprises without burning down the building.

The most practical way to read a purchase agreement is to ask four questions, repeatedly, like a detective who doesn't get bored.

First: *What moves, exactly?* Not "the business"—the shares, the assets, the contracts, the permits, the employees, the IP. If it's an asset deal, which liabilities are assumed and which are excluded? If it's a stock deal, what liabilities are sitting inside the entity, and how is the buyer protecting itself?

Second: *When does risk transfer?* There is a difference between economic risk (who suffers if performance drops) and legal risk (who owns the company). Signing and closing can be separate; interim operations can be constrained; some risks remain with the seller via covenants and closing conditions; some shift to the buyer via pre-closing access or financing obligations. The agreement maps that risk transfer with far more nuance than the parties' emotions do.

Third: *How does the math actually compute to dollars?* Follow the definitions, then the payment mechanics, then the schedules. Where are transaction expenses captured? How is debt calculated? Are there payoffs and releases? Who decides disputes over calculations? If the agreement includes a mechanism, it should also include a process.

If it doesn't, it's not a mechanism; it's an argument waiting for a calendar.

Fourth: *What has to be true for the deal to close, and who is responsible for making it true?* Consents, regulatory approvals, key employee agreements, payoff letters, deliverables, third-party actions. If a condition depends on a third party, that condition is not merely a condition; it's a risk allocation decision dressed as a checkbox.

Once you see the purchase agreement as a chassis and the ancillaries as functional components, the document becomes less intimidating and more legible. The point is not to memorize every clause. The point is to understand that M&A is a translation problem: translating commercial intent into enforceable, executable steps, with a pre-negotiated answer for what happens when reality disagrees. The definitive agreement is that translation. The accessories are the parts that make it run.

4.2 Price Adjustments and Earnouts: Moving Targets

A purchase price in M&A is a little like the "price" on a home listing: it tells you what kind of conversation you're having, but it isn't the number that clears your bank account. In deals, the headline figure is often *Enterprise Value*—the value of the business *as an operating machine* before you decide who gets the cash in the bank, who pays off the debt, and whether yesterday's bills are somebody else's problem. The number that gets wired at closing is usually smaller, messier, and surprisingly sensitive to definitions that look harmless until you run them through a spreadsheet.

The reason is not that bankers can't count. It's that businesses don't sit still. Cash moves every day. Vendors get paid. Customers pay late. Inventory gets built and shipped. Employees earn bonuses that haven't hit payroll yet. Meanwhile, the buyer is trying to buy a business that feels "normal" on Day 1, not a business that has been quietly stripped of fuel, tools, and spare parts in the final weeks before handoff. Price adjustments exist to make "normal" enforceable.

Two Prices Living in the Same Deal

Most private-company acquisitions implicitly contain two different prices.

One price answers: "What is the business worth *in the abstract*?" That's Enterprise Value (EV). It's the number you hear first because it's a clean sentence: "$200 million for the company." EV is meant to be capital-structure-neutral: it pretends the company has no debt and no extra cash, because those are financing choices and timing artifacts, not operating performance.

The other price answers: "What check gets written, to whom, at closing?" That's *Equity Value* and ultimately the *Closing Payment*. It starts with EV and then gets pushed around by a set of agreed-upon levers: net debt, cash, transaction expenses, and usually working capital. The definitive agreement turns those levers into a calculation engine with deadlines, disputes, and real dollars attached.

The classic shorthand for this engine is *cash-free, debt-free*. That phrase gets tossed around like it's self-explanatory. It isn't. "Cash-free" can mean "the seller keeps cash above a certain amount" or "all cash is swept and reduces the purchase price" or "cash stays in the company and effectively increases equity value." "Debt-free" can mean "bank debt only" or "every obligation that smells like financing, including leases, unpaid bonuses, and change-of-control payments." The economic result lives inside the definitions.

The Cash-Free/Debt-Free Bargain (And Why It's Never as Clean as It Sounds)

At its best, a cash-free/debt-free construct is an elegant handshake: the buyer pays for the operations; the seller settles the capital structure. Buyer gets a company that can run; seller gets paid for the value created. Everyone goes home.

At its worst, it becomes a game of labeling. Is that customer prepayment "debt"? Is that deferred revenue really a liability the buyer is inheriting? Are accrued vacation and bonuses treated as operating working capital or as debt-like items that reduce the price? Is an earnout obligation the buyer is agreeing to pay later effectively part of the price (and therefore not "debt"), or is it treated like indebtedness that reduces the closing payment? Each of these questions can be worth enough money to change someone's tone in a negotiation.

There's also an emotional trap here: people hear "debt" and think "bad." But the agreement isn't moralizing; it's allocating. Some obligations belong with the operations and should stay with the company (trade payables, accrued expenses that recur normally). Others are really financing choices that the seller benefited from (a dividend recap, shareholder loans, unusual related-party payables). The buyer's goal is not to punish debt; it's to avoid paying twice for the same value.

The Waterfall: Turning Enterprise Value into a Closing Wire

The purchase agreement's economic heart is often a waterfall calculation. It's the moment where the deal stops being narrative and becomes arithmetic.

Below is a stylized waterfall. Numbers vary wildly by deal, but the shape is recognizably the same: start with Enterprise Value, subtract debt-like items and transaction expenses, adjust for working capital relative to a target, then land on the closing payment—the check that actually gets written.

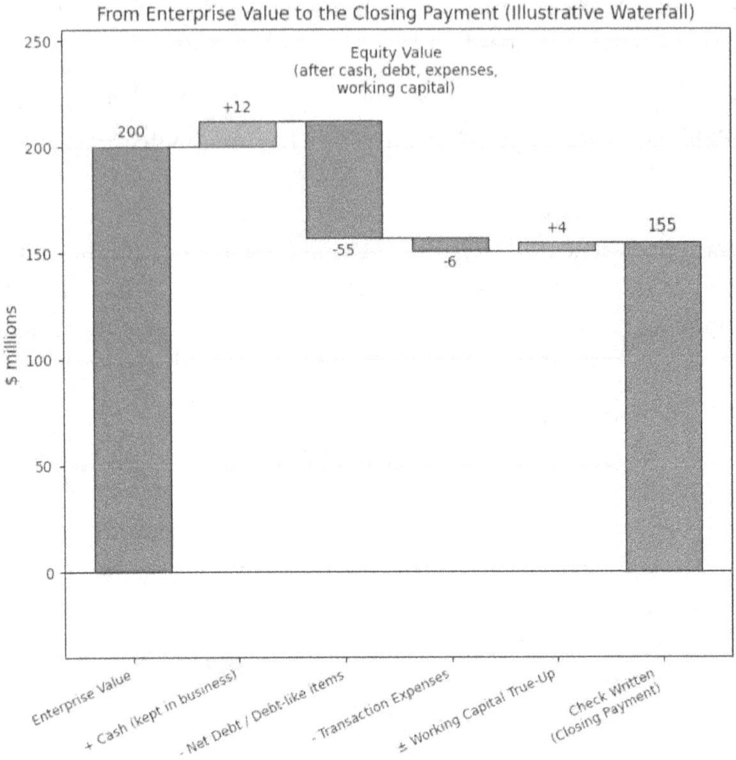

That last bar is the number that matters emotionally: it's the check, the wire, the escrow funding, the payoff letters getting satisfied in real time. But the earlier bars are where the negotiation actually lives. The parties can be "aligned" on enterprise value and still be miles apart on debt-like items, expense treatment, and working capital. That gap is why great deal teams obsess over definitions and schedules. They're not being pedantic; they're defending the math.

Working Capital: The Gas in the Tank

A business can look profitable and still be starving. That's what working capital captures.

Net working capital (NWC) is usually current assets minus current liabilities, with a definition that is always more specific than that simple formula. In plain English: it measures whether the business has enough short-term resources (cash collected from customers, inventory ready to sell) to cover its short-term obligations (paying vendors, payroll accruals, taxes due soon). If the buyer buys a company and immediately has to inject cash just to keep the lights on, the buyer didn't buy what they thought they bought.

The working capital adjustment tries to solve that problem by creating a target: the seller promises to deliver the company with "normal" working capital at closing. If closing NWC is above the target, the buyer pays more (because the buyer is receiving extra fuel). If closing NWC is below the target, the buyer pays less (because the seller effectively took fuel out of the tank before handing over the keys).

The beauty of the concept is also its danger. "Normal" is not a fact; it's a negotiation.

The target is the deal

The target NWC (sometimes called a *peg*) is often computed as an average of historical NWC over some period, sometimes adjusted for seasonality or growth. But averages lie in predictable ways. A seasonal business can have a working capital profile that looks healthy in the wrong month and terrifying in the right one. A fast-growing company can have permanently rising receivables and inventory needs. A business that recently changed billing terms might have a step-change in NWC that a backward-looking average ignores.

So the target becomes a proxy for a deeper fight: is the buyer paying for the business the seller *was* or the business the seller *says it is becoming*? Sellers tend to argue for a lower target (easier to exceed, increasing proceeds). Buyers tend to argue for a higher target (harder to fall short, reducing the chance of post-close cash injections). The negotiations can feel like accounting, but the stakes are operational: who funds Day 1.

Definitions: the silent steering wheel

The most common source of working capital fights is not the arithmetic. It's the definition of what counts.

Consider a few items that routinely get debated:

- *Deferred revenue:* If customers paid in advance, is that a liability that reduces working capital (and therefore reduces the price), or is it simply part of the business model the buyer is buying?
- *Accrued bonuses and vacation:* Are these treated as part of normal operations (working capital), or as debt-like obligations that should reduce the closing payment dollar-for-dollar?
- *Taxes payable:* Are they included in NWC or treated separately under tax covenants?
- *Related-party balances:* Are they real operating items or effectively shareholder financing that should be stripped out?

The purchase agreement usually answers these questions with a combination of (i) the NWC definition, (ii) an illustrative schedule (an example calculation), and (iii) consistency requirements (e.g., prepared in accordance with GAAP, applied consistently with past practices). Those words sound reassuring. In a dispute, they can point in different directions. GAAP has choices; "consistent" can preserve bad habits; examples can be incomplete. The best drafting doesn't just say "GAAP" and hope. It specifies inclusion/exclusion and gives an example that is not merely illustrative but genuinely clarifying.

The closing statement: a small document that can start a big war

Mechanically, working capital adjustments often work through a *closing statement* process. The buyer (or sometimes the seller) prepares a statement shortly after closing showing the computed NWC, net debt, cash, and other adjustment items. The other side has a window to review and object. If there's a dispute, it goes to a neutral accountant or expert for determination.

This is where time becomes a weapon. If the buyer controls the books post-close, the seller worries about "creative" classification. If the seller controls the books pre-close, the buyer worries about aggressive collections, slow-pay vendors, inventory dumps, or other last-minute moves that inflate earnings but drain working capital. The adjustment is supposed to neutralize those incentives. The process has to be tight enough to keep it from becoming a second negotiation after the wire has already hit.

A subtle but crucial drafting choice here is whether the neutral is acting as an *expert* or an *arbitrator*. The words sound like synonyms in normal conversation; in legal reality they can change procedure, scope, and judicial review. Many agreements try to keep this narrow: the neutral decides accounting and calculation issues, not legal interpretation. That boundary should be explicit, because when a working capital dispute goes off the rails, it's often because the parties are fighting about what the definition *means* rather than what the numbers *are*.

Net Debt and Transaction Expenses: The ``Who Pays for Yesterday" Problem

Working capital is about delivering a business that can operate. Net debt is about delivering a business without a financing hangover. Transaction expenses are about delivering a business without deal-cost landmines.

Net debt sounds straightforward: total debt minus cash. But the definition of debt in a purchase agreement tends to be much broader than a bank loan. It often includes capital leases, unpaid interest, prepayment penalties, letters of credit (sometimes), seller notes, and other obligations that function like financing. The logic is consistent: if an obligation is really a way the seller financed the business (or extracted value), the buyer doesn't want to pay for the business *and* assume the financing.

Transaction expenses are the invoices that appear because the deal happened: banker fees, legal bills, accounting fees, diligence consultants, sometimes success bonuses. The reason they're in the purchase agreement at all is simple: these costs need to be paid, and if they sit in accounts payable at closing, they effectively reduce working capital and become the buyer's problem unless the agreement strips them out. Buyers generally want transaction expenses paid by the sellers or deducted from the purchase price. Sellers, of course, want the company to pay as many of them as possible. The negotiation is not about fairness; it's about accurately pricing the deal as understood. If the parties agreed on EV assuming the seller pays deal costs, then the agreement needs to make that true.

Earnouts: Buying Tomorrow Without Overpaying Today

Working capital adjustments and net debt are about getting the business you thought you were buying at closing. Earnouts are about the business you hope you're buying afterward.

An *earnout* is contingent consideration: the seller receives additional purchase price later if the business hits agreed performance targets (revenue, EBITDA, ARR, gross margin, product milestones, regulatory approval, customer retention). Earnouts show up when there's a valuation gap that isn't just posturing. The seller genuinely believes growth is around the corner; the buyer genuinely believes it might not happen. The earnout says: fine—price the base case now, and pay extra if the upside materializes.

Earnouts also act like a lie detector for forecasts. A seller who is confident in the pipeline may accept an earnout as a way to protect value. A seller who is *merely hopeful* will resist, because earnouts are harder to collect than people think.

The difficulty is that earnouts are not purely mathematical. They are behavioral contracts in disguise.

Why earnouts get ugly

Once the deal closes, the buyer controls the steering wheel. That creates immediate tension: the seller wants the buyer to run the business in a way that maximizes the earnout metric; the buyer wants to run the business in a way that maximizes overall value, including integration synergies, product strategy, and risk management. Those two paths can coincide—but they don't have to.

Even when everyone is acting in good faith, measurement discretion can swamp the intent. If the earnout is based on EBITDA, then accounting policies, allocations of corporate overhead, hiring decisions, and revenue recognition can all change the number. If it's based on revenue, pricing strategy and bundling can change the number. If it's based on ARR, contract terms and renewals can change the number. An earnout metric is a magnet for every operational decision that affects it.

This is why earnout drafting often looks obsessive: defined terms, examples, exclusions, and covenants about operating the business. The parties are trying to anticipate how ordinary business decisions can accidentally (or deliberately) become earnout sabotage.

Choosing the metric: pick your poison carefully

Earnout metrics tend to trade off simplicity, manipulability, and relevance.

Revenue is simple and usually observable, but can be inflated by discounting, channel stuffing, or selling unprofitable deals. EBITDA tracks profitability, but is famously sensitive to allocations, management decisions, and accounting choices. Product milestones can be objective, but can be gamed by defining milestones loosely or by deprioritizing resources post-close. Customer retention metrics can align with real value, but can be affected by integration disruption that isn't the seller's fault.

The most seller-friendly earnout is the one the seller can measure without begging for information. The most buyer-friendly earnout is the one that can't be "won" by burning the future to juice the metric. A workable earnout sits in the uncomfortable middle: specific enough to be measurable, thoughtful enough to reflect value, and constrained enough to avoid turning the post-close period into trench warfare.

Operating covenants: the difference between a bet and a brawl

Earnouts often include covenants about how the buyer will operate the business: maintain separate books, use consistent accounting policies, refrain from actions primarily intended to reduce the earnout, use commercially reasonable efforts to achieve milestones, or avoid diverting sales away from the earnout business.

Every one of these phrases can trigger an argument. "Commercially reasonable efforts" sounds comforting until you ask: reasonable to whom, compared to what? "Not primarily intended" can become a motive trial. "Consistent accounting" can preserve bad practices or conflict with the buyer's enterprise policies.

Good earnout drafting does two things simultaneously: it *limits discretion* in the areas that matter most to the metric, and it *admits reality* that the buyer must run an integrated company. The art is to specify guardrails, not handcuffs.

Information rights: earnouts run on trust, but they settle on documents

If a seller is going to wait years for contingent consideration, they need visibility. Information rights can include periodic financial statements, metric reporting, access to underlying calculations, and the ability to discuss results with finance personnel. Buyers are understandably wary of opening their books too wide, especially after integrating the target. The compromise is often a reporting package tailored to the earnout metrics, with confidentiality obligations and limits on burden.

Without defined information rights, a seller's earnout claim can degrade into a vibe: "You're doing something to the numbers." With defined information rights, disputes can become concrete: line items, allocation policies, and specific accounting treatments.

Dispute resolution: deciding who decides

Earnouts, like purchase price adjustments, often route disputes to a neutral accountant or expert for calculation issues. But earnouts can also involve legal interpretation (what does "new product" mean?) and operational covenants (did the buyer use required efforts?). Many agreements split the baby: accounting disputes to an accountant; covenant disputes to courts (often Delaware, sometimes another chosen forum). The goal is to keep math disputes out of full litigation while preserving judicial oversight for behavior disputes.

This design matters because earnout disputes often feel personal. Sellers can perceive underpayment as betrayal; buyers can perceive claims as seller's remorse. A clear dispute path can prevent those emotions from hijacking the process.

Practical Drafting Knobs that Change Real Economics

A few structural choices repeatedly determine whether these mechanisms do their job or become expensive entertainment:

Interim covenants vs. adjustments: If the agreement tightly constrains pre-close operations, the buyer may be less reliant on post-close adjustments. If covenants are loose, the adjustment mechanism becomes the primary protection. The two should be drafted to complement each other, not fight each other.

Timing: Short deadlines force issues to surface quickly but can be unrealistic for complex businesses. Long deadlines give breathing room but invite drift and memory loss. The right answer depends on how fast reliable numbers can be produced.

Thresholds and materiality: De minimis thresholds can prevent nuisance disputes. But if thresholds are too high, they create a safe zone for value leakage.

Examples and exhibits: An illustrative calculation attached as an exhibit can be worth more than pages of prose. It reduces ambiguity by showing how the parties intend the definitions to behave with real numbers.

These aren't accessories. They are the hidden switches that decide whether the buyer gets a fair adjustment or a headache, whether the seller gets a fair earnout or a fight.

The Real Point: Preventing Value from Teleporting

Price adjustments and earnouts can feel like accounting games. They're not. They're anti-teleportation devices.

Without them, value can move between signing and closing (or between closing and the earnout end date) in ways that are hard to see and easy to regret. A seller can accelerate collections, delay payables, underinvest in inventory, or push costs out of the period. A buyer can integrate in a way that buries revenue, allocate overhead aggressively, or change accounting policies to shift profitability. Most people are not scheming; they're optimizing for their own incentives. The contract's job is to make those incentives less destructive.

The cleanest deals are not the ones with the fewest mechanisms. They're the ones where the mechanisms are simple enough to run, specific enough to measure, and boring enough that nobody wants to litigate them. When a purchase agreement achieves that kind of boring, it's doing something sophisticated: it's turning a moving target into a number you can wire.

4.3 Reps, Warranties, and the RWI Revolution

Every acquisition has a moment where the buyer realizes a simple truth: you can diligence a company for months and still not *know* it. You can interview management, tear apart the general ledger, read every customer contract, and still be surprised by something that was always there—a missing consent, a tax position taken years ago, a salesperson who promised the moon in writing, a contractor who never assigned IP. Businesses are too complicated, and human memory is too selective, for certainty.

Representations and warranties ("reps" and "warranties") are the contractual answer to that uncertainty. They are not warm reassurances. They are bets written in legal ink: the seller states that specific facts about the business are true, and if a statement is wrong, someone pays. The phrase "snapshot" is apt, but incomplete. A snapshot doesn't just capture what's there; it also freezes blame. It turns "we didn't know" into "who bears the loss anyway."

Reps and Warranties as a Risk-Allocation Engine

Reps and warranties look like information—lists of truths about the company. Their deeper purpose is economic. They decide which side eats which surprises.

A buyer is usually comfortable with *known* risks if they are priced in or structurally addressed. The problem is the unknown unknowns: the ugly thing nobody spotted, or the benign-looking thing that later proves expensive. Reps and warranties create a map of those unknowns by dividing the world into three buckets:

- Things the seller is affirmatively saying are true (and will stand behind).
- Things the seller is admitting as exceptions in disclosure schedules.
- Things neither side wants to talk about but that still might exist (and will be allocated by whatever is left of the deal's remedies).

That structure gives reps their peculiar power. They don't just punish dishonesty. They make honesty *payable*. If the seller discloses an exception, the buyer can react: negotiate price, demand a covenant, require a consent, carve out the risk, insist on a special indemnity, or walk. If the seller doesn't disclose and the rep is wrong, the seller has (in theory) just agreed to write a check later.

The Difference Between ``We Think'' and ``We Promise''

In normal business life, companies speak in probability. "To our knowledge." "We believe." "We haven't seen anything." M&A reps force language to pick a side.

A rep that says "The Company is in compliance with all laws" is a hard promise (and often too broad to be believable without qualifiers). A rep that says "To the Company's Knowledge, the Company is in compliance with all laws" is a softer promise: it's really a statement about what the seller knows, not the objective truth. A rep that says "except where the failure would not be material" introduces another layer: now you're negotiating not only facts, but *materiality*.

This is where reps become a kind of contract-grade lie detector. Sellers push for knowledge qualifiers and materiality qualifiers because they reduce exposure. Buyers push for unqualified statements where surprises would be genuinely value-destructive. Neither side is being philosophical. They're pricing risk.

The Cast of Characters: The Reps Buyers Fight For

The reps that generate the most heat are the ones tied to liabilities that can be enormous, long-dated, or hard to detect in diligence.

Financial statements reps can be deceptively narrow or deceptively broad. In private deals, it's common to see reps about whether financials were prepared in accordance with GAAP and whether they fairly present the financial condition of the company. Those phrases sound interchangeable; they aren't. "GAAP" can be a compliance test; "fairly present" can invite arguments about judgment calls, estimates, reserves, and classification. A seller might concede one and resist the other not because they're hiding fraud, but because they don't want an accounting debate to become a damages claim.

Taxes reps are the quiet monsters. Tax liabilities can survive leadership changes, ignore your integration plan, and show up with penalties. Sellers may represent that returns have been filed and taxes paid, that no audits are pending, that withholding has been handled properly, that there are no tax liens, that the company hasn't participated in certain listed transactions. Buyers care because tax authorities do not negotiate the way private counterparties do. They assess. They collect. They garnish. If there is one category of risk that buyers reliably want backed by real dollars, it's taxes.

IP reps can be existential in technology and brand-driven businesses. The buyer wants comfort that the company owns or has the right to use its IP, that key employees and contractors assigned inventions, that the company isn't infringing others, and that it has taken reasonable steps to protect trade secrets. Diligence can catch obvious problems, but chain-of-title issues hide in old files and forgotten contractor arrangements. IP reps turn those ghosts into someone else's financial responsibility.

Compliance and regulated matters (privacy, data security, export controls, anti-corruption, healthcare regulations, environmental laws) are increasingly central even in "ordinary" middle-market deals, because the cost of being wrong can be lopsided. A small company can accidentally violate a law that carries enterprise-threatening penalties. Buyers often insist on specific reps tailored to the target's risk profile. Sellers often insist those reps be knowledge-qualified or materiality-qualified. The negotiation is rarely about abstract virtue; it's about whether the buyer is willing to accept a hidden compliance landmine.

Contracts reps look boring until you remember contracts are where revenue lives. Buyers want to know that material contracts are valid, enforceable, and not in default; that there are no undisclosed side letters; that the company hasn't given away most-favored-nation pricing in a way that will crush margins; that customers can't terminate at will due to a change of control. A sophisticated buyer will often accept that some contracts are messy. What they won't accept is being told the contracts are clean and discovering the mess after closing.

Disclosure Schedules: Where the Truth Actually Resides

If reps are the seller's promises, disclosure schedules are the seller's confession—in a way that can still be strategic.

Schedules serve two jobs at once. First, they provide lists (contracts, litigation, employees, benefits plans, permits, IP registrations). Second, they create exceptions to the reps. A rep might say "The Company has no pending litigation" and the schedule says "except the following." This is how a seller avoids lying while still providing a clean representation structure.

But schedules also create a different kind of risk: the risk of *incomplete* disclosure that is technically compliant but practically misleading. A schedule can list the existence of a problem without conveying its seriousness. It can bury important facts in attachments. It can comply with the letter of a disclosure obligation while violating the spirit of what the buyer thought it was buying.

Great buyers read schedules like investigative journalists. They look for patterns: repeated customer complaints, recurring regulatory themes, contract clauses that appear in multiple forms, a surprisingly large number of "immaterial" disputes with the same counterparty. Great sellers treat schedules like a last diligence pass. They want to avoid the worst possible outcome: a post-close claim where the seller says, "We disclosed it," and the buyer says, "Not like *that*." Even if the seller wins, they've just bought themselves years of misery.

Bring-Downs, Survival, and the Difference Between Signing and Closing Truth

Reps are often made at signing and then "brought down" at closing—meaning the seller repeats that they are still true as of closing, sometimes subject to specified materiality thresholds. This matters because a lot can change between signing and closing: a customer churns, an employee quits, a regulator sends a letter, a data incident happens. If reps are brought down broadly, the seller takes the risk that normal life produces a technical breach and gives the buyer an out. If reps are brought down only in limited ways, the buyer takes the risk that something meaningful changes and they're still forced to close.

Then comes *survival*: how long reps remain actionable after closing. Survival turns reps into a time-limited warranty policy. It answers: how long does the buyer get to discover a problem and make a claim? A short survival period is seller-friendly (less tail risk). A longer one is buyer-friendly (more time for latent issues to surface). Certain reps—taxes, fundamental authority, capitalization, title—often survive longer because the risks they cover don't respect short timelines.

Survival is also an admission about what diligence can't do. Most problems don't present themselves on Day 1. They show up when a customer dispute becomes litigation, when a regulator shows up, when a tax audit happens, when an ex-employee asserts they own code. Survival is the deal's answer to that latency.

The Traditional Model: Reps Backed by the Seller's Balance Sheet

Historically, the enforcement mechanism for reps was straightforward: if the rep was wrong and the buyer suffered losses, the seller would indemnify (pay back) the buyer, often using an escrow as security.

In that model, the seller's post-close exposure loomed large. Sellers had to worry that months after closing, they'd be dragged back into a fight. Buyers had to worry that even if they were right, collecting would be slow, contentious, or impossible. An escrow helped, but escrows are finite, and severe breaches can exceed them.

This model also shaped negotiations in a particular way: sellers fought hard to narrow reps and limit remedies because they were personally on the hook; buyers fought hard to keep them broad because the seller was the only realistic source of recovery. The negotiation energy often went into limiting the *seller's* pain rather than simply pricing the *business's* risk.

The RWI Revolution: When the Insurer Becomes the Seller's Pocket

Representation & Warranty Insurance (RWI) changed the emotional physics of that negotiation.

RWI is an insurance policy that covers certain losses arising from breaches of reps and warranties. Conceptually, it replaces (or heavily reduces) the seller's post-close liability with an insurer's balance sheet. The buyer still cares about the reps; in fact, the reps become even more central, because the policy typically only covers breaches of *written* reps in the definitive agreement. But the source of payment shifts. Instead of the buyer suing the seller for indemnification, the buyer often makes a claim under the policy.

That shift is why RWI feels revolutionary: it turns a potentially personal, relationship-damaging dispute into something more procedural. The seller gets a cleaner exit. The buyer gets a potentially deeper pocket. Deals can close with smaller escrows and less seller holdback. Competitive auctions can move faster because sellers can offer buyers a smoother "clean exit" package and still give them a meaningful backstop.

Of course, insurance never just removes risk. It *moves* it, reshapes it, and adds friction in new places.

What RWI Actually Changes at the Term Sheet Level (Even If Nobody Says It Out Loud)

RWI's biggest impact is not the existence of a policy. It's the domino effect on deal terms.

Sellers in RWI deals often push for minimal indemnification obligations—sometimes only for "fundamental" reps, covenants, and certain special items. Buyers accept that because they expect the policy to do the heavy lifting. Escrows shrink because the insurer is the primary remedy source. Survival periods can be aligned to the policy term rather than the seller's appetite for lingering liability. The negotiation becomes less about whether the seller will be ruined by a claim and more about whether the *insurer* will cover it.

This re-centers the entire discussion. Instead of asking, "How much risk can the seller stomach?" the parties ask, "What risk will the insurer underwrite, and at what price?"

The Catch: RWI Is Written in Exclusions

RWI has a reputation among non-practitioners as a magic shield. It isn't. It is a contract with its own definitions, conditions, and exclusions, and those exclusions often track the exact areas buyers worry about most.

Known issues (things discovered in diligence) are generally excluded. If the diligence process turns up a problem and the buyer still proceeds, RWI is not designed to be a subsidy for that choice. Certain categories (like underfunded pensions, some environmental matters, certain cyber incidents, forward-looking statements, or purchase price adjustments) may be excluded or treated specially. If a deal has a particular hot spot—say, a specific tax risk—the parties might pursue separate insurance (a tax policy) or a special indemnity, because RWI won't cover it cleanly.

This is where drafting and diligence become inseparable from insurance. The buyer's recovery depends on the reps. The reps are negotiated. The reps are scheduled. The schedules reflect diligence. The insurer underwrites based on diligence. The whole system is a loop, not a stack of independent parts.

Underwriting: The Strange Moment Where the Buyer Pays Someone to Question Them

RWI underwriting is one of the more surreal rituals in modern dealmaking. The buyer (often) is buying the policy, paying the premium, and then sitting across from an underwriter whose job is to find reasons not to pay.

Underwriters will review diligence reports, ask detailed questions about red flags, and probe areas that look light. They are not performing full diligence for you; they are assessing whether your diligence is good enough for them to take the risk. If your diligence is thin, the policy may be more expensive, include more exclusions, or simply be unavailable on acceptable terms.

This creates a quiet discipline in the market: RWI can raise the floor on diligence quality. Not because underwriters are altruists, but because they have a direct financial interest in whether your process actually uncovered what it should have uncovered.

Claims: The Dispute Doesn't Disappear, It Changes Counterparties

RWI also changes who you argue with after closing.

In a traditional indemnity claim, the buyer and seller fight about (i) whether there was a breach, (ii) whether the buyer suffered a covered loss, (iii) whether the loss is limited by baskets, caps, and survival, and (iv) whether the buyer mitigated. In an RWI claim, the buyer may fight about similar issues—but with the insurer. And insurers bring their own instincts: coverage analysis, policy language, causation arguments, and procedural requirements.

That can be a feature or a bug. It can be a feature because it reduces the chance the seller's leadership team spends the next year in depositions. It can be a bug because insurers don't pay claims out of goodwill or reputation management; they pay under contract, and their contract is designed to be interpreted tightly.

The best mental model is not "RWI eliminates disputes." It's "RWI relocates disputes to a different arena with different rules."

Anti-Reliance and the Hard Boundary of the Written Reps

One of the most consequential modern drafting moves around reps is the anti-reliance framework: language that says the buyer is relying only on the reps and warranties expressly set forth in the agreement, and not on other statements (management presentations, CIMs, diligence discussions, emails, optimistic commentary).

This language can feel insulting in the moment. It's also a precision tool. It draws a boundary around what is actionable as fraud or misrepresentation and forces the parties to put important statements into the actual rep set if they want them to be enforceable.

RWI tends to reinforce this discipline. Insurers generally cover breaches of *contractual* reps. If a buyer wants insurance-backed protection, it must make sure the relevant facts are represented in the agreement and not merely implied in a slide deck.

The result is a subtle cultural shift in dealmaking: the definitive agreement becomes the center of gravity for truth. The oral story matters less. The written rep matters more. The schedule matters most.

A Practical Way to Read Reps (Without Drowning in Them)

Reps are long. They are also repetitive across deals. The trick is to read them with a filter that keeps your attention on what moves money.

When a rep is qualified by "materiality," ask: material to whom, and measured how? When a rep is qualified by "knowledge," ask: whose knowledge, and is knowledge actual, constructive, or inquiry-based? When a rep is heavily scheduled, ask: is the schedule a real disclosure or a data dump? When a rep is broad and unqualified, ask: is the buyer actually planning to rely on it, or is it there as negotiating leverage to get a better schedule?

And always ask the most useful question in this entire area: if this rep is wrong, what's the remedy and who pays? Without a remedy, a rep is just a sentence. With a remedy, it's a pricing instrument.

Why the Revolution Matters

RWI didn't make sellers more honest or buyers more trusting. It did something more practical: it made certain deals easier to close.

By shifting a large portion of rep breach liability away from the seller and onto an insurer, RWI can reduce the post-close shadow hanging over sellers and reduce the buyer's fear that a valid claim will be uncollectible. It can help auctions run cleaner, allow management teams to roll equity without keeping one foot in old disputes, and let financial sponsors return capital without keeping a reserve for unknown claims.

But it also demands a higher standard of precision. Because once you move from "we'll work it out with the seller" to "we'll make a claim under a policy," every word starts to behave like it's going to be read by someone trained to say no. Reps become sharper. Schedules become more important. Diligence becomes more than a comfort exercise. The agreement stops being merely a record of what everyone believes and becomes what the insurer will recognize as reality.

That's the paradox of the RWI era: the more insurance you buy, the more the contract itself matters.

4.4 Indemnification: The ``Money Back'' Guarantee

A deal closes, the wire lands, everyone exhales—and the company immediately goes back to being a company. Customers complain, employees quit, regulators regulate, invoices arrive late, and sometimes the past, which had been politely kept off the Zoom calls, shows up at the front door with paperwork.

Indemnification is the part of the purchase agreement that decides what happens *then*. It is the money-back guarantee on the seller's factual promises and certain covenant obligations, written with the cold specificity of a warranty card: what counts as a covered defect, how long the coverage lasts, how big the deductible is, the maximum payout, and the address where claims must be mailed (with the right exhibits, by the right date, to the right person). People talk about indemnification like it's "legal." It's not. It's economics with deadlines.

The most important mental shift is this: indemnification isn't primarily about punishing bad behavior. It's about allocating the cost of surprises. If a representation about taxes was wrong, the question is not whether someone feels embarrassed. The question is whether the buyer eats the tax bill, the seller pays it, or an insurer pays it, and whether the answer changes depending on the size and timing of the bill.

Indemnification is a System, Not a Clause

Indemnification provisions are often described as a "basket" and a "cap" with an escrow. That description is like calling a car "an engine and four wheels." True, but missing the point. An indemnity regime is a system with interlocking parts:

- *Scope:* what losses are covered (rep breaches, covenant breaches, specific items).
- *Process:* how a claim is made, when it must be made, and who can make it.
- *Limits:* baskets, caps, and carve-outs.
- *Duration:* survival periods.
- *Security:* escrows/holds, setoff rights, guarantees, or insurance.

Change any one of these, and the deal's risk profile changes, even if the headline purchase price doesn't move a penny.

What Counts as a ``Loss" (And Why Parties Argue About the Word)

Before talking about baskets and caps, it helps to understand what an indemnity claim is actually claiming.

A buyer doesn't sue for "a breached rep" in the abstract. They sue for *Losses* arising from that breach. The agreement typically defines Losses in excruciating detail: damages, costs, expenses, penalties, interest, reasonable attorneys' fees, amounts paid to third parties, and sometimes consequential or special damages (either included or excluded), sometimes diminution in value (either allowed or explicitly disallowed), sometimes multiples of damages (often excluded), sometimes the buyer's internal costs (often excluded).

This definition is where the indemnity regime either becomes a practical remedy or an academic one. If "Losses" exclude the kinds of harm that actually occur in the buyer's world, the reps start to feel ornamental. If "Losses" include every imaginable category, the seller sees a blank check. The negotiated middle reflects what kind of deal this is: a friendly transaction with a clean exit, or a risk-heavy acquisition where the buyer insists on a wide recovery net.

A particularly contentious subtopic is whether the buyer can claim *diminution in value*—essentially the reduced value of the company due to the breach—instead of (or in addition to) out-of-pocket costs. Sellers hate this because it can look like the buyer is trying to claw back part of the purchase price based on a valuation argument. Buyers like it because many breaches don't cause a neat invoice; they cause a long-term reduction in profitability or growth. Agreements often channel this tension by allowing diminution claims only in certain circumstances, or by requiring losses to be third-party damages or actual amounts paid.

Survival Periods: The Warranty Clock Starts at Closing

Reps don't live forever. Survival periods are the clock on the wall.

A standard approach is that "general" reps survive for a relatively short period (often measured in months rather than years), while certain categories survive longer: fundamental reps (authority, capitalization, title), taxes, and sometimes employment/benefits or environmental. The logic is intuitive: the more catastrophic or slow-to-surface the risk, the longer the buyer wants to keep the seller on the hook.

Survival is not just about fairness; it is about latency. A tax audit doesn't ask your permission to be delayed. A wage-and-hour claim can take time to crystallize. A missing software license might not be discovered until the buyer tries to scale usage. If the survival period expires before a reasonable buyer could discover and assert the claim, the rep becomes a nice story with no ending.

Survival periods also interact with claim notice mechanics in ways that can feel like legal trivia until you miss the deadline. Many agreements require that a claim be noticed before the survival period ends, sometimes with specific detail. A buyer who sends a vague email saying "we may have tax issues" might not have preserved anything. A seller who receives a poorly drafted notice might later argue it didn't qualify as a claim. This is not hypothetical; it's where deals go to litigation.

Baskets: The Deductible That Prevents Death by a Thousand Paper Cuts

If survival is the clock, the basket is the deductible.

A basket is a dollar threshold of losses the buyer must absorb before the seller owes anything for general rep breaches (and sometimes certain covenant breaches). The basket exists because buyers can find *something* wrong in almost any company if they look hard enough. Without a basket, the post-close relationship can devolve into constant, petty claims: $12,000 here, $8,500 there, each one expensive to argue about relative to its size.

The basket forces the buyer to save up grievances until they are meaningful. It is, frankly, a way to keep adults from litigating like toddlers.

But baskets come in two main flavors, and they behave very differently once the threshold is crossed:

- *Deductible (or "true deductible") basket:* the seller pays only the amount of losses *above* the basket.

- *Tipping (or "first dollar") basket:* once losses exceed the basket, the seller pays *all* losses, sometimes from dollar one (or from a lower sub-threshold).

A deductible basket is closer to how insurance works: you eat the first slice, then coverage kicks in. A tipping basket is more dramatic: it's like a trapdoor. Once you hit the threshold, the seller is suddenly paying for losses that were previously the buyer's problem.

Sellers prefer deductible baskets because they keep exposure incremental. Buyers prefer tipping baskets because they preserve the ability to recover meaningful amounts once the buyer proves the existence of a real problem.

Caps: The Seller's Maximum Exposure

If the basket is the deductible, the cap is the policy limit.

The cap sets the maximum aggregate amount the seller can be required to pay for certain categories of indemnifiable losses (typically general rep breaches). The cap is often expressed as a percentage of the purchase price. In a seller-friendly deal, the cap might be small, especially when RWI is in play. In a buyer-friendly deal without insurance, the cap might be larger.

Caps are not always uniform. Many agreements have a lower cap for general reps and a higher (sometimes uncapped) exposure for certain items: fundamental reps, fraud, and sometimes specific indemnities. This is the agreement admitting that not all breaches are morally or economically equivalent. A breach about a minor contract default is not the same as lying about ownership of shares.

Caps are also where parties express their view of what a deal is supposed to feel like. Some sellers want the deal to be final, not an installment plan with clawbacks. Some buyers want a period where the seller remains financially invested in the accuracy of their story. The cap is where those philosophies become arithmetic.

A Picture Worth a Thousand Negotiation Emails: Basket, Cap, and the Indemnifiable Range

The basket and cap are easier to understand visually than verbally. The claim amount is the water level. The basket is the dry zone where the seller pays nothing. The cap is the rim where the seller's liability ends. Everything in between is the indemnifiable range.

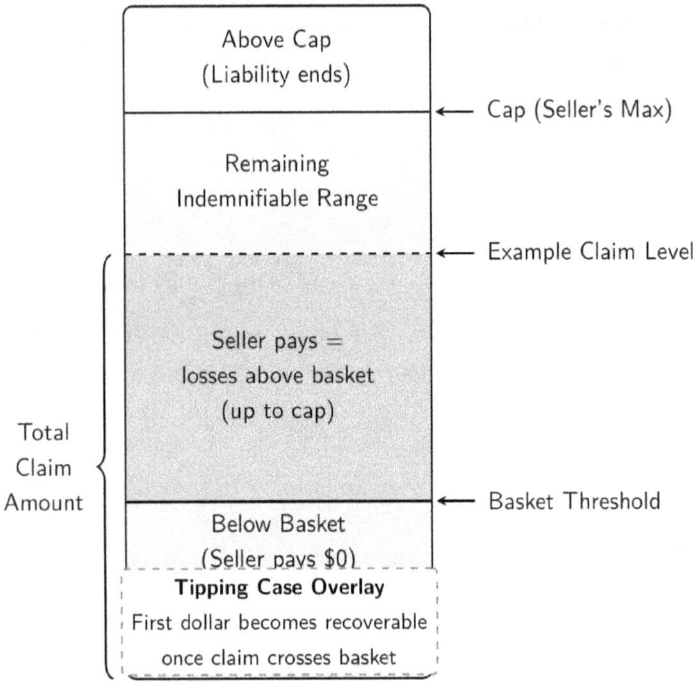

This picture also explains why buyers and sellers obsess over one deceptively small drafting question: is the basket deductible or tipping? In the diagram, a deductible basket means the seller pays only the blue area *above* the basket line (up to the cap). A tipping basket means once the claim level crosses the basket line, the seller may pay starting from the very bottom of the bucket (or near it), effectively turning the gray zone into recoverable territory after the fact.

Escrow: The Security Deposit That Makes Indemnity Real

An indemnity promise is only as good as the ability to collect.

In many private deals, the buyer's practical recovery source is an escrow funded at closing. The escrow is a portion of the purchase price held by a third-party agent under an escrow agreement. It exists for one reason: to prevent the buyer from having to chase sellers across states, cap tables, and tax returns to collect on valid claims.

Escrows also change behavior. A seller who knows money is locked up is often more willing to compromise on claims to release the remainder. A buyer who knows recovery is real is often more comfortable closing even when diligence reveals some manageable messiness. Escrow is not just security; it's a pressure valve.

But escrow introduces its own negotiations: amount, duration, release mechanics, and whether claims can be satisfied exclusively from escrow or also from sellers directly. Sellers push for a small escrow and a short release schedule. Buyers push for a larger escrow and a longer tail. When RWI is used, escrows often shrink dramatically, because the insurer becomes the primary backstop and the escrow is reserved for a narrow band of seller-retained risks.

The escrow agreement itself is a companion document with teeth. Escrow agents are not referees with discretion. They are machines that process instructions. If the release mechanics require a final non-appealable judgment, funds can be trapped for years. If they allow unilateral notices with minimal support, funds can be frozen too easily. The right design is a negotiation about leverage disguised as administrative procedure.

Carve-Outs: The Places Where the Cap and Basket Don't Apply

No indemnity regime is complete without exceptions. These carve-outs are where the deal reveals what truly scares the buyer.

Common carve-outs include:

- *Fundamental reps:* authority, capitalization, ownership, sometimes title to assets.
- *Taxes:* sometimes subject to separate caps/survival, sometimes not.
- *Fraud:* often uncapped and with longer survival (and sometimes expressly preserved regardless of other limitations).
- *Special indemnities:* a known, specific risk identified in diligence (a particular lawsuit, a specific tax exposure, a regulatory matter) with its own bespoke rules.

Carve-outs matter because they prevent the seller from buying peace too cheaply. A seller may accept a low general cap if they are still meaningfully exposed to the handful of risks that could actually blow up the buyer's investment thesis.

Carve-outs also create a drafting challenge: if every rep is treated as "fundamental," then nothing is. Precision is the point. When everything is special, the agreement becomes a mood rather than a mechanism.

Mitigation, Offsets, and Insurance Proceeds: The ``Don't Double Dip" Negotiation

Indemnification is not only about whether a loss is covered; it's about what happens around the loss.

Most agreements require the indemnified party to mitigate losses—to take reasonable steps to reduce damages. This sounds like common sense, and it can be weaponized. A seller may argue the buyer failed to mitigate by making a business decision that increased losses. A buyer may argue mitigation shouldn't require them to distort operations post-close just to protect the seller. Many agreements land on a reasonableness standard, but even that word can turn into a trial if the stakes are high enough.

Offsets are another pressure point. If the buyer makes a claim, can the seller reduce the claim by pointing to insurance proceeds the buyer received? If the buyer recovers from a third party (say, a customer settlement), does that reduce what the seller owes? Sellers push hard for these provisions to avoid paying for losses that are already reimbursed. Buyers push for flexibility because insurance recovery can be uncertain, delayed, or partial.

RWI makes this even more interesting. In insured deals, the buyer may have a policy that covers rep breaches, and the seller may still have some indemnity obligations (for example, for certain excluded matters, covenants, or fraud). Agreements can specify how insurance interacts with seller indemnity to prevent the buyer from collecting twice—while also avoiding a situation where the seller hides behind insurance and the insurer hides behind the seller.

Third-Party Claims: Where the Process Really Matters

Direct claims ("our working capital was misstated") are one thing. Third-party claims are where indemnification becomes a procedural sport.

A third-party claim might be a lawsuit, a demand letter, a regulatory inquiry, a tax assessment, an employment claim. The indemnity section typically lays out who controls the defense, who selects counsel, who has approval rights over settlement, and how information must be shared. Buyers often want control because the claim affects their newly acquired business and reputation. Sellers often want control (or at least meaningful consent rights) because it's their money funding the defense and settlement.

This is not just about ego. Control affects outcomes. A buyer might settle quickly to avoid distraction. A seller might fight to avoid precedent or to reduce payout. A buyer might accept non-monetary settlement terms that complicate operations. A seller might refuse a reasonable settlement to gamble on saving money. The agreement's control mechanics determine whose instincts win.

Why Indemnification Still Matters in an RWI World

With RWI, it's tempting to treat seller indemnity as an artifact, like fax machines and three-hole punchers. But indemnification remains structurally important even in heavily insured deals.

First, insurance has exclusions. Someone still bears those excluded risks, and the purchase agreement decides who. Second, insurance is a process. It can take time to pay, and claims can be contested. A seller may still provide a small escrow to cover the deductible/retention under the policy or certain limited seller obligations. Third, fraud and certain fundamental matters often remain with the seller regardless of insurance posture. The buyer needs a clean, enforceable indemnity framework for those categories.

In other words, the money-back guarantee doesn't vanish. It just changes which counterparty writes the check, and which set of definitions you end up litigating.

The Practical Reality: Indemnification is Negotiated Peace

Indemnification is where a deal's optimism meets its realism. Everyone wants to believe the business is as represented, the numbers are solid, and diligence found what mattered. Indemnification is the agreement admitting that even smart people miss things.

A well-built indemnity regime does something quietly valuable: it makes post-close conflict *finite*. It sets thresholds so the buyer doesn't nickel-and-dime. It sets caps so the seller can sleep. It sets survival so both sides know when the past becomes the past. And it funds a security deposit (escrow, holdback, or insurance) so the promises aren't performative.

When it's drafted well, indemnification is rarely tested, because its existence shapes behavior and discourages games. When it's drafted poorly, it becomes the afterlife of the deal—a second negotiation conducted through claim notices, spreadsheet battles, and lawyers' calendars. The difference is not luck. It's whether the deal team treated the indemnity provisions as real economics rather than boilerplate.

Chapter 5

Managing the Path to Closing

You have signed the definitive agreement and popped the champagne, but you don't own the keys yet. Welcome to the 'interim period'-the most nerve-wracking phase of M&A where the deal exists in legal limbo. This chapter walks you through the delicate art of keeping the business steady while satisfying the conditions required to release the wires, ensuring you don't trip over a covenant or regulatory hurdle just inches from the finish line.

5.1 Operating in Limbo: Interim Covenants

Between signing and closing, the seller is still holding the keys—but the buyer has already paid for the right to complain if you scratch the paint. That awkward overlap is what interim covenants are built for: they are the contract's way of saying, "Keep the business steady, don't get cute, and if something big has to happen, ask first." It sounds simple until you remember what running a company actually looks like: customers threaten to leave, equipment breaks, competitors cut prices, and talented employees decide they've always dreamed of moving to Denver *right now*.

Interim covenants turn those everyday surprises into a governance problem. Not a moral one. Not a "be reasonable" one. A governance one: who is allowed to decide, under what standards, and with what paper trail. The discipline they impose can feel unnatural to an operator, because operators are rewarded for speed. Interim covenants reward *process*. If you treat them as mere legal fine print, you'll discover their real function the hard way: they are a pre-built litigation script for a deal that is no longer a handshake and not yet a marriage.

The ``ordinary course" promise: the deceptively short sentence that runs your calendar

The headline interim covenant is usually some variation of: the seller will operate the business "in the ordinary course consistent with past practice" between signing and closing. That phrase is short enough to fit in a margin, and powerful enough to determine whether the buyer has to close.

The key words are not "ordinary" or "course." The key words are *consistent with past practice*. "Ordinary" in an operator's mouth can mean "what a sensible person would do given the circumstances." "Consistent with past practice" is colder. It means the company's own history becomes the measuring stick. If the business has historically handled a situation one way, and you handle it another way—even if your new approach is arguably smarter—you've created an argument that you breached.

That's why this covenant is so slippery during disruption. In a crisis, managers want to do whatever works. Interim covenants want you to do what you *usually* do, or to get consent for doing something else. The covenant is not trying to make you a bad CEO; it's trying to stop you from becoming a different CEO once someone else is buying the results.

Why buyers care: value can leak without anyone ``doing anything wrong"

A buyer signs an acquisition agreement based on a snapshot: the diligence materials, the budget, the quality of earnings, the customer roster, the headcount plan. Between signing and closing, that snapshot becomes stale on day one. The buyer's fear is not only outright value stripping (the cartoon villain version). It's more subtle: a well-intentioned seller making a dozen "small" calls that are each defensible, but collectively change the business the buyer thought it was purchasing.

Common "leakage" patterns rarely sound dramatic in isolation:

- A nervous sales team pulls forward discounts to keep bookings up, leaving the next quarter hollow.
- A plant manager delays preventive maintenance to protect short-term cash, and failures spike later.
- A CEO tries to be kind and accelerates retention bonuses for key employees, quietly changing the economics of the deal.
- A founder decides that, since they're selling anyway, now is the perfect time to settle an old lawsuit "for peace," even if the settlement is richer than market.

None of these is necessarily theft. They are all *deal-relevant*. Interim covenants exist because value can leak through decisions that look, from inside the company, like normal leadership.

Two buckets: (1) ``run it normally,'' and (2) ``don't do these things without permission''

Interim covenants typically come in two flavors that work together:

1. *Affirmative operating covenants*, led by "ordinary course," requiring the seller to keep running the business in a familiar pattern.
2. *Negative covenants* (often called "negative interim covenants"), a list of specific actions the seller agrees not to take without the buyer's consent.

Think of ordinary course as the *default setting*. Negative covenants are the *alarm system*. One says, "keep doing what you've been doing." The other says, "and definitely don't do *these* things unless the buyer says yes." The list of negative covenants is where deals start to feel operationally real, because it's where a corporate attorney's imagination collides with a CFO's calendar.

What shows up on the ``don't do this" list (and why each item exists)

Negative covenants look repetitive across deals because the underlying risks are repetitive. The buyer is trying to prevent the seller from (a) changing the economics, (b) changing the risk profile, or (c) changing what the buyer is actually acquiring.

Here are the usual suspects, translated into plain English and linked to the buyer's fear:

Negative covenant category	*What the buyer is actually worried about*
Dividends, distributions, share buybacks	Cash leaves the business (or the capital structure changes) before the buyer takes over.
Incurring debt, granting liens, refinancing	The buyer ends up paying for a different balance sheet than the one priced, or inherits a web of liens that complicate payoff and closing.
Issuing equity, options, changing comp plans	The ownership picture shifts, dilution happens, or employee incentives get re-cut in a way that changes retention cost post-close.
Capital expenditures above a threshold	Cash gets spent on assets the buyer didn't choose, or spending is deferred so the buyer inherits a maintenance backlog.
Entering/amending material contracts (esp. long-term or non-ordinary)	The seller locks in pricing, exclusivity, or service levels that hurt margin or constrain future strategy.
Hiring/firing senior people; changing wages/benefits	The org chart gets reshaped, severance obligations spike, or key talent exits because a decision was rushed.
Settling litigation or tax matters above a threshold	The seller buys certainty with the buyer's money (or changes the risk allocation embedded in indemnities).
Selling assets or making acquisitions	The buyer's "what am I buying?" question gets a moving answer.
Changing accounting policies or reserving practices	Reported performance and working-capital mechanics can be nudged without changing underlying cash.

The art is not memorizing the list. The art is understanding the *logic*: if an action can change value, timing, or risk in a way that cannot be easily unwound at closing, it tends to be consent-gated.

Consent isn't a nuisance; it's the operating system

A mature seller treats buyer consent like a workflow, not an interruption. That requires two mental shifts.

First: consent requests should be written as if a stranger will read them later. Because a stranger might. If the deal blows up, your email chain becomes evidence. "Need approval ASAP thx" reads differently in court than "Requesting consent under Section 5.1(b)(ix) for a one-time customer concession consistent with prior renewal practice; expected revenue impact $X; no change to term; no exclusivity."

Second: consent is easiest to obtain when the buyer believes you have control of the business. A seller who looks panicked, disorganized, or opportunistic gets slow-walked. A seller who presents a crisp choice set—what is happening, why it matters, what the alternatives are, and what happens if you do nothing—gets decisions.

A practical cadence that works in real deals is a standing weekly "interim covenant" call with a tight agenda: open consent requests, upcoming actions likely to trigger consent, and a quick review of KPIs that tend to produce surprises (cash collections, customer churn, headcount changes, capex, large payables). The point is not bureaucracy for its own sake; it's to avoid the Friday-at-6:30pm email where you ask for consent to a decision that was effectively already made.

The budget trap: ``It's in the budget" is helpful, but not a force field

Many agreements tie ordinary course to an operating plan or budget delivered during diligence. That sounds comforting: if it's in the budget, it must be allowed. In practice, budgets create their own traps.

A budget can support a consent request ("we're executing the plan you already underwrote"), but it rarely answers everything:

- Budgets are often annual; deals sign mid-year; the relevant decisions are weekly.
- Budgets assume things happen smoothly; real businesses don't.
- A line item labeled "Professional Fees" doesn't tell you whether settling a lawsuit for $2.5 million is "in budget."

When "in the budget" matters most is in shaping the buyer's expectations. If you can credibly show that an action is consistent with what the buyer priced (or at least saw coming), consent becomes easier. But don't treat the budget like a hall pass. Treat it like supporting evidence.

Materiality: the word that turns a small fire into a legal one

Interim covenants often use qualifiers like "in all material respects," "Material Contracts," or dollar thresholds for specific actions. Sellers love these qualifiers. Buyers fear them. The tension is inevitable: the seller wants to run the business without calling the buyer for every stapler; the buyer wants to avoid death by a thousand "immaterial" cuts.

Two practical points make materiality less mystical:

Materiality is often about reversibility. A $200,000 expense might be financially immaterial in a $300 million business, but if it's tied to a three-year exclusive distribution agreement, it can be strategically material. Courts and buyers alike tend to treat long-lived, hard-to-unwind decisions as more important than one-time noise.

Materiality is often about pattern, not size. One off-cycle retention bonus might be explained. Three off-cycle bonuses, a new commission plan, and a management carve-out start to look like a deliberate reshaping of the company. Buyers don't just audit *numbers*; they audit *behavior*.

The cleanest way to reduce fights is to define thresholds thoughtfully and to separate recurring decisions from one-time events. A capex threshold that forces consent for every $40,000 server purchase will annoy everyone and still fail to catch the strategic decisions. A better approach is typically a combination: per-item thresholds, aggregate thresholds over a period, and explicit treatment for certain categories (like customer concessions or executive compensation) that are strategically sensitive even when the dollars aren't huge.

The ``back-door exit": covenant compliance can matter more than a MAC

Everyone hears about the scary "Material Adverse Change/Effect" concept (the idea that a sufficiently bad deterioration can allow a buyer to walk). In practice, the ordinary course covenant can be a more realistic tripwire, because it's not asking whether the business got worse; it's asking whether the seller *followed the rules* while it got worse.

That difference matters. A downturn might not be severe enough (or sufficiently company-specific) to qualify as a deal-breaking adverse change. But if management reacts by making sweeping operational changes without consent—closing sites, furloughing staff, slashing capex, rewriting customer terms—the buyer may argue it doesn't have to close because the seller failed a separate condition: covenant compliance. In other words, even if the storm isn't a legal hurricane, steering the ship off the agreed channel can still breach.

The practical lesson is not "be timid." It's "negotiate for flexibility *before* you need it, and document consent *when* you need it."

Carve-outs: where realism sneaks into legal drafting

Good interim covenant packages contain carve-outs that acknowledge reality. Some are standard:

- Actions required by law or by a governmental authority.

- Actions taken in response to emergencies, subject to notice.
- Actions taken with buyer consent (with the buyer agreeing not to "unreasonably withhold, condition, or delay" consent in some cases).

But the most valuable carve-outs are tailored to the actual business model. A seasonal business needs the ability to hire and spend in bursts. A business with volatile commodity inputs needs flexibility in hedging. A company mid-way through a product launch needs latitude for marketing and inventory decisions that might look "non-ordinary" if you only glance at last quarter.

Sellers often discover too late that "required by law" is narrower than it feels. "Permitted by law" is not "required by law." If a regulation gives you discretion, the carve-out may not help. Likewise, "emergency" can turn into a debate if you didn't treat it like an emergency at the time (no board notes, no documented advice, no rapid escalation). If you want freedom to act, it's better to negotiate it explicitly than to hope a judge shares your definition of urgency.

Managing the business without triggering a consent avalanche

The job is to keep the business healthy *and* keep the deal alive. That requires triage.

First, identify the "high-friction" zones early. These are the areas where normal business operations and deal protections naturally collide: pricing concessions, retention packages, capex catch-up, collections strategy, vendor renegotiations, and anything involving the CEO or CFO's signature. If you know a union contract renegotiation is coming up in six weeks, don't wait five weeks to tell the buyer.

Second, pre-clear categories, not just single actions. The most efficient consent is a mini playbook agreed in writing: "You can offer up to X% discount on renewals within these parameters," or "You can replace departing managers within these comp bands," or "You can execute capex items already in the approved list up to an aggregate amount."

Buyers often prefer this too, because it reduces the chance they will accidentally delay a decision that later hurts performance.

Third, treat buyer consent like a service-level obligation. Sellers sometimes assume the buyer is dragging its feet out of malice. Often it's just process: internal approvals, lender involvement, or investment committee sensitivities. Make it easy for the buyer to say yes quickly by giving them what their own organization will ask for: quantified impact, alternatives, and a clear statement of why the action is consistent with past practice or necessary for the business.

Employee moves: the most emotional interim covenant battleground

Nothing tests "ordinary course" like people decisions. Deals make employees anxious, and anxious employees make expensive choices. Interim covenants frequently restrict:

- hiring or firing above a certain level,
- salary increases beyond a percentage,
- retention, transaction, or stay bonuses,
- new equity grants or option repricings,
- changes to benefit plans.

The buyer's logic is straightforward: compensation changes alter the economics and can create obligations that survive closing. The seller's reality is equally straightforward: the business can't hit its numbers if key people leave.

The smart way to navigate this is to treat retention as a joint problem with a joint budget. If the seller unilaterally starts throwing money at departures, the buyer may later argue it's value leakage. If the buyer refuses all retention spend, performance can sag, and the buyer ends up protecting the deal mechanics at the expense of the asset.

A practical technique is to propose a *retention framework* rather than one-off exceptions:

- identify key roles (not just names),
- define triggers (resignation risk, competitor offers, project milestones),
- set maximum amounts and forms (cash vs. bonus vs. post-close employment agreements),
- clarify who approves and how quickly.

It's easier for a buyer to approve a controlled program than a series of "we're going to lose Alex by Friday" emergencies.

Customer and vendor contracts: where ``ordinary" can secretly mean ``short-term"

Many negative covenants focus on "Material Contracts" and on contracts outside the ordinary course. The trap is that a contract can be ordinary in *type* but extraordinary in *terms*. Renewing a customer annually is ordinary; renewing them for five years at a discount with an exclusivity clause is not. Extending a vendor agreement is ordinary; adding a minimum purchase commitment that changes working capital dynamics is not.

Sellers should get in the habit of scanning proposed changes for three deal-sensitive features:

1. *Term*: longer terms lock in economics beyond closing.
2. *Exclusivity / non-competes*: these can constrain strategy and attract antitrust attention.
3. *Commitments*: minimum volumes, take-or-pay, or service-level penalties can move risk from the counterparty to the company.

If any of those features are present, assume you need consent (or at least a written "no consent required" confirmation) even if the dollar value feels modest.

Capex and maintenance: the silent fight over tomorrow's headaches

Capex restrictions exist to prevent the seller from spending the buyer's money on pet projects. But the mirror-image risk is more common: sellers *underinvest* to make short-term metrics look better. Deferred maintenance is a classic pre-closing value leak because it creates a clean P&L today and a messy plant tomorrow.

Buyers know this, which is why sophisticated interim covenants sometimes contain affirmative obligations to maintain assets, keep insurance in place, and preserve relationships with key suppliers and customers. Even when the contract doesn't say it explicitly, a seller that slashes maintenance, slows safety spend, or stops replacing worn equipment is taking a legal and commercial risk: it can degrade performance and hand the buyer a narrative that the seller didn't operate "as usual."

The operational answer is boring and effective: keep your maintenance schedule intact, and if you need to deviate, document why and seek consent when deviation would be visible in post-close performance. A one-page capex tracker shared with the buyer weekly can do more to preserve trust than any eloquent email after the fact.

Financial reporting and ``no surprises": the covenant you wish you had even when you don't

Many agreements require the seller to provide periodic financial statements, notice of certain events, and reasonable access to information. Sellers sometimes view these as intrusive. Buyers view them as early-warning systems. The closer you get to closing, the more the buyer's focus shifts from diligence (what is this company?) to confirmation (is it still that company?).

A seller trying to keep optionality may be tempted to slow-roll bad news. That instinct is understandable and usually counterproductive. Bad news that arrives late looks worse, triggers more buyer process, and increases the risk that a solvable problem turns into a closing delay. Interim covenants, at their best, encourage a healthier habit: surface issues early, frame them accurately, and propose solutions that preserve the business while respecting consent rights.

How to write a consent request that gets answered

A consent request is a miniature investment memo. It should be short, concrete, and drafted so that the buyer can forward it internally without rewriting it. A reliable structure is:

- *What you want to do* (one sentence).
- *Why* (two or three sentences, with context).
- *Impact* (quantified where possible: revenue, margin, cash, headcount, timing).
- *Alternatives considered* (so the buyer sees you didn't jump to the most aggressive option).
- *Time sensitivity* (real deadline, not manufactured urgency).
- *Ask* (clear "consent requested under [cite the covenant]" language).

When sellers skip "impact" and "alternatives," the buyer has to do that work itself. That's when the request stalls, not because the buyer is malicious, but because you've handed them an incomplete file.

The seller's internal control: appoint a ``covenant gatekeeper"

The most common interim covenant breach is not a dramatic CEO decision. It's a mid-level leader doing something routine—signing a contract, approving a pay raise, committing to a spend—without realizing it tripped a consent threshold. Deals fall apart in the gaps between legal language and operational muscle memory.

A practical control is to name one person (often the CFO, GC, or a dedicated deal lead) as the covenant gatekeeper. The gatekeeper's job is not to micromanage. It's to be the single chokepoint for actions that could be consent-gated, with a lightweight routing rule:

- anything above $X,
- anything with a term beyond Y months,
- anything involving executives, equity, or unusual compensation,
- anything that changes standard customer terms,
- anything that looks like "we don't usually do this."

This is less about preventing bad acts and more about preventing accidental ones. Interim covenants punish accidents just as effectively as intent.

The buyer's behavior matters too: consent rights are not a license to freeze the business

Buyers sometimes overuse consent as leverage, slowing approvals to renegotiate price or to force concessions elsewhere. That approach can backfire. A target that cannot make timely decisions can miss sales, lose employees, and disappoint customers—damaging the very asset the buyer is trying to protect.

That's why many agreements build in some friction against buyer abuse: standards like "not unreasonably withhold" consent, requirements to respond within a specified time, or carve-outs that allow the seller to act in emergencies with prompt notice. Even when those protections are absent, a buyer that wants the deal to close tends to behave like a partner in the interim period, not a traffic cop. The interim period is long enough for resentment to grow, and short enough for it to be costly.

The north star: preserve the business, preserve the bargain, preserve the record

Interim covenants are not asking the seller to stop managing. They are asking the seller to manage in a way that preserves three things at once:

The business: keep customers, talent, operations, and financial discipline intact.

The bargain: don't change the economics or risk profile the buyer agreed to without giving the buyer a say.

The record: create a clean paper trail showing that when unusual decisions were necessary, they were handled transparently, consistently, and with the right approvals.

Do that, and the interim period becomes what it should be: not a limbo, but a controlled handoff where the company keeps winning on Tuesday while the lawyers quietly make sure the deal still deserves to close on Friday.

5.2 The Escape Hatches: Closing Conditions and MACs

A signed deal feels like certainty until you read the verbs. The purchase agreement doesn't say "Buyer will buy." It says "Buyer will buy *if*..." and then hands you a long, carefully engineered list of reasons the parties can delay, refuse, or litigate. Those "ifs" are closing conditions: the contract's bouncer rope. If you meet them, you get in. If you don't, the other side gets a legal story for why the party with the money is allowed to keep it.

Closing conditions are also the part of a deal where businesspeople discover a lawyer's favorite magic trick: turning mushy judgment calls into binary switches. "The company is doing fine" becomes "the reps are true as of closing to the specified standard." "Nothing crazy happened" becomes "no Material Adverse Effect has occurred." "We kept our promises" becomes "covenants performed in all material respects." It's not romance; it's engineering.

Conditions precedent: the checklist that can stop a $500 million wire

A closing condition (often called a "condition precedent") is simply something that must be true—or must have happened—before a party is obligated to close. Think of it as the difference between a reservation and a nonrefundable ticket. A signed merger agreement is binding, but it's usually not an unconditional promise to close on a date certain. It's a promise to close when the conditions are satisfied (or waived).

There's an important tone difference here. When someone says, "We have the right to walk away," that sounds like a loophole. When the agreement says, "We are not required to close unless these conditions are satisfied," it's not a loophole at all. It's the deal. Closing conditions define what "successfully getting to close" actually means in legal terms, not in vibes.

Why conditions exist: nobody wants to buy yesterday's company

Between signing and closing, time passes and life happens. The target can lose a key customer, get sued, face a regulatory delay, or have its CFO quit in protest. The buyer wants protection against paying the original price for a different company. The seller wants protection against the buyer inventing reasons to retrade once the seller is publicly committed.

So the contract tries to balance two fears:

- *Buyer's fear*: "I'm being forced to buy a business that changed materially."

- *Seller's fear*: "The buyer will use technicalities to renegotiate price or walk."

Closing conditions are where those fears negotiate in plain sight.

The common architecture: bring-down, covenants, approvals, and ``no MAE"

Most deals end up with a familiar skeleton of conditions, even if the muscle and scars are different.

Bring-down of representations and warranties. Representations ("reps") are factual statements about the business: financials are prepared properly, contracts are valid, compliance is in order, taxes are paid, no undisclosed liabilities, and so on. The buyer typically requires that these statements be true at signing and also true at closing—or at least true to a negotiated standard.

That "standard" is where the drama hides. Reps might have to be true:

- *in all respects* (seller hates this),
- *in all material respects* (everyone argues about what "material" means),
- *except where failures would not have a Material Adverse Effect* (a higher bar, often buyer-friendly because it concentrates disputes into a single definition),
- or through a "materiality scrape" mechanic that changes how materiality is measured for indemnification (a separate fight, but it seeps into closing logic).

Covenant compliance. Covenants are promises about behavior between signing and closing: operate in the ordinary course, don't do certain things without consent, provide access, cooperate on filings, and so on. The typical closing condition says the target has complied with its covenants "in all material respects." That sounds forgiving until you realize what it enables: a buyer that can't prove a Material Adverse Effect may still argue it can refuse to close because the seller breached a covenant that is itself a condition.

Regulatory approvals and third-party consents. Some conditions are pure gating items: the government has to clear the deal, the landlord has to consent, the key customer has to sign off, the lender has to fund. These aren't about whether the business is "good." They're about whether the transaction is *allowed* and *possible* to consummate.

No Material Adverse Effect ("no MAE" / "no MAC"). This is the famous one: a condition that, broadly speaking, says the buyer doesn't have to close if a sufficiently bad event has occurred. It's the clause that gets the headlines, and it's the clause that most rarely does the work people think it does.

The real-world feel: conditions are not symmetric, even when they look symmetric

Agreements often have closing conditions for both sides: buyer conditions and seller conditions. Superficially symmetrical drafting can hide a very asymmetrical reality, because the buyer usually has more conditions and more discretion. Sellers typically require things like: the buyer's reps are true, the buyer has performed its covenants, financing is available (sometimes not), and required approvals are obtained. Buyers require those too—and add a list of conditions tied to the target's business and risk.

That asymmetry is not always "unfair." It reflects who is bearing post-close risk. Once the buyer closes, it owns the mess. The seller (especially in a cash-out deal) wants to be done. So the buyer pushes risk back into the pre-close period via conditions. The seller pushes back by narrowing, clarifying, or pricing that risk.

Bring-down mechanics: you're not re-selling the company at closing, but you are re-certifying it

A bring-down condition is essentially a re-confirmation: "Those facts you promised were true at signing are still true at closing." It's common for the seller to deliver a *bring-down certificate* signed by an officer at closing to that effect. That certificate is not mere ceremony; it's a personal signature on the idea that the seller is not trying to sneak a changed company through the gates.

What makes bring-downs interesting is how they translate a thousand small facts into one big question: *how wrong can the reps be and still require closing?* If reps have to be accurate "in all material respects," then a handful of minor inaccuracies shouldn't stop closing. If they have to be accurate "in all respects" for certain fundamental reps (like capitalization, authority, and title to shares), then even a small mistake can become a legal boulder.

Practically, bring-down disputes often start with something that sounds harmless: an updated schedule. The seller updates disclosure schedules between signing and closing to reflect new facts. Is that an allowed update that prevents surprise, or is it evidence that the reps are no longer true? The answer depends on drafting, but the emotional reality is simpler: schedule updates are where sellers confess change, and where buyers decide whether change is acceptable.

Covenant compliance: the ``back-door MAC" nobody brags about

Material Adverse Effect clauses get the spotlight because they're dramatic: the buyer tries to exit by saying the business got wrecked. Covenant compliance is quieter, and often more dangerous for sellers. If the seller promised to operate in the ordinary course and didn't, the buyer may have a cleaner contractual pathway to refuse closing: "You didn't satisfy a closing condition."

This matters because "ordinary course" is not a philosophy; it's a contract standard. If the covenant is written to past practice, then a seller can be in breach even while making rational decisions under stress. That can be infuriating to management ("We saved the company!") and legally irrelevant ("You saved it in a way the contract didn't allow without consent."). The covenant condition is the lever that turns a process failure into a closing failure.

The uncomfortable truth: many busted deals don't die because the business fell apart. They die because the parties lost trust, and covenant compliance gave the distrusting party a tool that looks principled in a termination notice.

Regulatory and third-party conditions: the difference between ``we want to close" and ``we're allowed to close"

Some conditions are moral judgments; approvals are physics. Antitrust clearance is not something you "feel good about" and proceed anyway. If you're in a regime that requires pre-closing approval, you are legally prohibited from closing until the waiting period expires or approval is granted. Same for many industry regulators. Same for certain contract consents that, if missing, would cause a key agreement to terminate upon closing.

These conditions also create the most common type of closing delay: not a fight about who's right, but a calendar problem. The business can be performing beautifully and the parties can be perfectly aligned, yet the deal can't close because a government agency's internal clock hasn't run out. That delay changes incentives. Sellers get impatient. Buyers keep diligence open longer. Employees get restless. Interim covenants start to bite harder because the limbo stretches.

The MAC/MAE clause: the monster under the bed that mostly just keeps kids from jumping

The Material Adverse Change/Effect clause (MAC/MAE) is often described as the buyer's escape hatch. That phrasing makes it sound like a big red button. In reality, it's more like a fire alarm behind glass: you can pull it, but you'd better be right, and everyone will be watching.

A MAC/MAE definition typically says something like: an event, change, effect, or circumstance that is materially adverse to the business, results of operations, or financial condition of the target (sometimes with specific carve-outs and exceptions). It is usually paired with a closing condition: no MAE has occurred since signing.

Why it's such a high bar is not mysterious. If MACs were easy to trigger, buyers could sign deals and then fish for reasons to exit whenever markets wobble or forecasts miss. Sellers would treat signed deals as meaningless, and the M&A market would gum up. So the legal system and deal practice have evolved toward making MAE a genuinely serious threshold: something that suggests a long-term impairment of the business, not a bad quarter.

Carve-outs: the MAC clause's hidden heart

Most of the MAC clause is actually the carve-outs: a list of things that *don't* count as a MAC. The carve-outs are where the parties decide which risks are simply part of being alive.

Common carve-outs include:

- changes in general economic conditions,
- changes in financial markets, interest rates, or credit availability,
- changes affecting the target's industry generally,
- acts of war, terrorism, natural disasters, pandemics (in many modern forms),

- changes in law or accounting rules,
- announcement effects (lost customers or employees because the deal became public),
- failure to meet projections (often carved out, with nuance about underlying causes).

Carve-outs often come with a *disproportionate effect* qualifier: if the target is hit disproportionately compared to other companies in its industry, then the carve-out may not protect the seller. That's the compromise between "macro stuff shouldn't be my problem" and "but what if macro stuff uniquely crushes your business?"

The practical meaning is that MAE fights are not just about whether something bad happened. They're about *categorizing* badness: is it general or company-specific, temporary or durable, foreseeable or shocking, and was the target hit harder than its peers?

Why buyers still insist on MAC conditions even though they rarely win

If MAEs are hard to prove, why do buyers care so much about having the condition? Because the condition is not only about winning in court. It's about bargaining power during a crisis.

When the world turns ugly between signing and closing, parties renegotiate. They might extend outside dates, re-trade price, add escrow, tighten covenants, or restructure financing. A buyer with a MAC condition has a credible argument that it has a contractual out. Even if that argument is uncertain, uncertainty itself creates leverage. Sellers know that litigating to force closing is expensive, distracting, and risky. The MAC clause becomes a shadow on the negotiating table.

In other words, the MAE is not just a legal standard; it's an *incentive structure*. It changes how hard each party will work to solve problems versus positioning for an exit.

Outside dates: the quiet escape hatch that actually gets used

If you want an escape hatch that gets used in the real world, pay attention to the "outside date" (also called a "termination date" or "drop-dead date"). This is the date after which either party can terminate if the deal hasn't closed, typically subject to exceptions (for example, a party can't terminate if its own breach caused the delay).

Outside dates are the contract's way of acknowledging that some deals don't die in a single dramatic moment. They die of delay. Regulatory reviews stretch. Financing wobbles. Third-party consents stall. The business performance drifts. The outside date forces a decision: either extend (often with concessions), litigate, or walk.

It's also the condition that gives time real monetary value. If the buyer can wait out the seller, the outside date can become a pressure tactic. If the seller can keep the business stable and show momentum, it can force the buyer to choose between closing and being the party that kills the deal.

Financing conditions: the line between ``strategic buyer certainty" and ``sponsored buyer reality"

Whether a deal has a financing condition is one of the most important risk-allocation choices, and it tends to correlate with the type of buyer.

A strategic buyer (a corporation buying for synergy) often resists making closing conditional on financing. They may have cash, revolvers, or investment-grade access to the market. Sellers push hard for this because it makes the deal feel like a true commitment: "Your ability to pay is your problem."

Private equity deals are more nuanced. The buyer entity is often a special-purpose vehicle that relies on committed debt and equity. Many modern sponsor deals are structured to give sellers high certainty without a classic "financing condition," but they may include other mechanisms that, in practice, price financing risk: reverse termination fees, limited remedies, equity commitment letters, and covenant packages requiring the sponsor to use "reasonable best efforts" to obtain the financing.

Even when a deal has "no financing condition," financing can still be an escape hatch if remedies are capped. If the buyer's worst-case outcome is writing a fee check rather than being forced to close, the economic reality is that the buyer has purchased an option: expensive, but still an option.

Remedies: when the escape hatch is less about permission and more about price

Two questions matter when a party wants out:

- *Can they terminate?* (conditions, covenants, MAE, outside date)
- *What happens if they're wrong?* (remedies: damages, specific performance, termination fees)

Specific performance—forcing a party to close—sounds like the seller's dream remedy. In practice, it's often limited, negotiated, and surrounded by conditions (for example, the debt financing must be available). Many agreements instead channel disputes into termination fees: a buyer pays a reverse termination fee if it fails to close under certain circumstances; a seller pays a break fee if it accepts a superior proposal (more common in public deals). Fees are not just punishment; they're a way of turning uncertain litigation into a predictable number.

This is where "escape hatch" becomes more precise. The contract may not give a party a clean right to walk away for free. It may give them the ability to walk away for a known price, like an airline change fee that hurts but doesn't ruin your life.

A practical way to think about deal certainty: four dials, not one

People talk about "deal certainty" as if it's a single trait. It's not. It's a set of dials the parties tune:

1. *How strict are the bring-down standards?* (How perfect must the reps be at closing?)
2. *How tight are the interim covenants, and how fast is consent?* (Can the business breathe without breaching?)
3. *How broad is the MAE definition, and how generous are the carve-outs?* (How much macro and idiosyncratic risk does the seller bear?)
4. *What are the remedies?* (If someone refuses to close, do you get a lawsuit, a fee, or a forced closing?)

Changing one dial often forces changes in others. A buyer that agrees to a narrow MAE and forgiving bring-down standards might demand tighter interim covenants. A seller that accepts heavy covenants might negotiate for quicker consent obligations. A sponsor buyer that can't tolerate uncapped damages might offer a larger reverse termination fee to compensate.

How deals actually break: the narrative usually matters as much as the clause

When a deal terminates, the termination notice will cite clauses. But the real driver is often a narrative that hardened over weeks: "the business is deteriorating and management is hiding it," or "the buyer got cold feet and is inventing issues," or "regulators are taking forever and we can't keep operating under these constraints."

Closing conditions and MAEs are the legal vocabulary for those narratives. That's why the most important work during the interim period is not only compliance; it's credibility. When a problem arises, the party that can tell the most coherent, well-documented, contract-consistent story gains leverage. The party whose story is a patchwork of late disclosures and improvised explanations usually loses, even if they are morally right.

The cleanest takeaway: treat closing like an exam with multiple sections

Signing feels like the finish line because it's loud: signatures, press releases, congrats texts. Closing is quieter and more clinical. It's an exam with multiple sections, graded by the agreement:

- Are the facts still true enough? (bring-down)
- Were the promises kept enough? (covenants)
- Did the world break the business badly enough? (MAE)
- Did the necessary strangers say yes? (approvals and consents)

Most deals close because the parties want them to close. The escape hatches matter because incentives change under stress. Closing conditions are the contract's way of ensuring that when stress arrives—and it usually does—the argument is about defined standards rather than raw power. The standards may still be contested, but they are at least written down, numbered, and attached to signatures.

5.3 Show Me the Money: Financing and Funds Flow

Deal people love to talk about purchase price the way sports fans talk about a score. "$425 million." "12x EBITDA." "All-cash." Clean numbers, clean bragging rights. Closing day is where those numbers get dragged into the real world, where money has to move through banks that don't care about your multiple, only your wiring instructions and cut-off times.

That translation layer—from headline price to *exact* dollars landing in *specific* accounts—is funds flow. It's not glamorous, but it's the moment the deal becomes irreversible. Contracts can be argued over. Wires, once sent, have a stubborn way of making everyone suddenly cooperative.

Financing is a promise; funds flow is choreography

A debt commitment letter is a lender's promise to provide financing if conditions are met. It can be long, negotiated, and expensive to obtain. It still isn't money. It's more like a movie studio greenlighting a film: you have a budget, you have conditions, and you have a lot of ways for things to get delayed if someone misses a cue.

Funds flow is the cue sheet. It's the operational plan that takes your Sources and Uses table (a tidy spreadsheet that says where money *comes from* and where it *goes*) and turns it into something banks can execute: who wires what, to whom, when, and under what release instruction.

When this goes well, closing feels anticlimactic: a few emails, a call, a set of PDFs, and then silence. When it goes poorly, it's a live financial autopsy: someone's payoff letter is wrong, a fee is missing, a wire is in the wrong currency, a lien release isn't authorized, and the parties discover that "close today" actually meant "close today if we're lucky."

Sources and Uses: the spreadsheet that lies by omission

Sources and Uses is the deal's basic conservation-of-cash equation. *Sources* are the dollars you can marshal: new bank debt, sponsor equity, rollover equity, seller notes, maybe cash on hand. *Uses* are where those dollars must go: purchase price paid to the seller, debt payoffs, transaction fees, escrow funding, sometimes a working capital injection.

The reason it lies (politely) is that it can look balanced while hiding timing and mechanics. A Sources and Uses tab can show "Payoff Existing Debt: $80 million" as a single line item. In reality, that payoff might be four separate facilities, each with its own per diem interest, breakage, notice requirements, and payoff instructions that change depending on the hour you close. It's the difference between writing "Dinner" on your calendar and actually cooking for twelve guests with allergies.

Funds flow forces honesty. It breaks the single line into the messy parts: principal, interest, prepayment fees, the agent bank's administrative fee, and the exact account numbers. It also forces you to face a question that makes CFOs sweat: *who is responsible if the payoff is short by $37,412.18 because interest accrued for one extra day?* In a well-run process, that question is answered before anyone sends a wire.

Why the paying agent is the hero nobody thanks

Most deals route funds through a paying agent (often the target's transfer agent in public deals, a bank, an escrow agent, or sometimes the seller's representative). The paying agent is the clearinghouse: money comes in from sources, then goes out to uses, often in a tightly controlled sequence.

The paying agent's value is not just convenience; it's control. It centralizes execution so that:

- the buyer isn't wiring ten different recipients directly,
- the seller isn't relying on the buyer to correctly calculate and pay third-party amounts,
- and counsel can condition releases on the satisfaction of closing conditions.

In practice, the paying agent is also the emotional firewall. When a seller calls asking, "Has the money hit yet?" the paying agent can answer without turning it into an accusation. On closing day, that neutrality is worth real money.

``Funds flow memorandum'': the single source of truth (and the best place to prevent a lawsuit)

The Funds Flow Memorandum (sometimes called a Flow of Funds, Closing Funds Memorandum, or Disbursement Schedule) is the authoritative document that says, in plain arithmetic: *these* funds are coming from *these* sources, and they will be disbursed to *those* recipients in *these* amounts, using *these* wiring instructions, on *this* date, pursuant to *these* authorizations.

It reads like a glorified invoice until you remember what it really does: it allocates blame ahead of time. If the memo is correct and followed, mistakes are rare and disputes are containable. If it's vague or "still being finalized" at the last minute, you are inviting chaos into the most expensive hour of the transaction.

A good Funds Flow Memorandum is also aggressively boring. No commentary, no negotiation history, no explanations that belong in an email. Just math, names, accounts, and references. Closing day is not when you want interpretive dance; it's when you want a checklist.

The hidden landmines: payoff letters, per diem, and the tyranny of cut-off times

The most common funds flow problems are not exotic. They're the kind you'd expect if you've ever tried to pay off a mortgage and learned that the payoff amount changes daily.

Payoff letters. A payoff letter is the lender's written statement of the amount needed to repay a facility in full as of a specified date, plus the wiring instructions and the lender's commitment to release liens once it receives good funds. Without a correct payoff letter, you can wire the money and still not get the lien releases you need. That's the nightmare scenario: the buyer paid, but the business is still encumbered.

Per diem interest. If the payoff is calculated as of Monday but closing slips to Tuesday, interest accrues. Someone must pay it. The funds flow memo should show whether the payoff is "good through" a date, what the per diem is, and who is responsible for the delta.

Wire cut-off times. Banks have cut-off times for same-day wires, and they are not sympathetic to your closing call. Time zones matter. So do compliance reviews. A wire that needs manual approval at 4:45pm can become a next-day problem, which can become a contractual dispute if the agreement requires closing by a certain time. The simplest fix is procedural: plan for morning funding even if the lawyers think the call will be in the afternoon.

Escrow: money you pay now so you can argue later

Escrow is a deal's way of admitting, politely, that not all disagreements can be resolved before closing. The buyer wants protection for indemnification claims or purchase price adjustments. The seller wants the majority of proceeds now. Escrow splits the difference: a portion of the purchase price is held by an escrow agent and released later based on the contract's rules.

In funds flow terms, escrow is just another use of funds with its own wiring instructions. In deal psychology terms, escrow is a pressure valve. It reduces the temperature of closing day disputes because the buyer doesn't have to threaten to withhold the entire purchase price to secure leverage; it has a defined pot of money set aside.

The escrow agent's release mechanics are where everyone gets serious. Typically, the escrow agent will not release funds based on vibes. It will release based on a joint instruction, or a final non-appealable order, or a unilateral instruction after a notice period if the other side doesn't object. Those mechanics can turn a future argument into a manageable process—or into trench warfare. Funds flow is where you make sure escrow is actually funded exactly as the contract contemplates, not "close enough."

Rollover equity: when the seller is both payee and investor

Rollover equity is the most conceptually confusing line item for non-specialists because it's the same person wearing two hats. Part of the seller's consideration is rolled into equity in the post-close company. In human terms: the seller is taking chips off the table, but leaving some on the felt.

In funds flow terms, rollover is often not a "wire to seller." It can be:

- a reduction in the cash paid at closing, paired with the issuance of equity,
- or a cash payment to the seller followed immediately by a subscription back into the buyer entity,
- or a netting arrangement where the paying agent sends less cash and counsel documents the equity issuance.

Each method has different tax, timing, and documentation implications, but the operational risk is always the same: if you don't map the rollover clearly, someone will think they are owed cash when they are actually owed stock, or vice versa. That confusion does not improve with adrenaline.

Transaction fees: the swarm around the honey jar

Closing day attracts fees the way light attracts moths. Legal fees, accounting fees, banker success fees, lien release fees, filing fees, escrow fees, sometimes bonus payouts triggered by the transaction. Some are paid by the buyer, some by the seller, some by the company, and some are negotiated in ways that feel arbitrary until you remember that every dollar paid by the company is a dollar not paid to the seller (and, in a cash-free/debt-free construct, can change the economic bargain).

Funds flow is where you prevent the most awkward phone call in M&A: the banker asking why their invoice isn't included on the disbursement schedule *after* the wires have gone out. The right discipline is to treat fees like debt payoffs: collect invoices early, confirm amounts, confirm payees, confirm wiring instructions, and confirm who is responsible. Vague obligations like "Seller shall pay all transaction expenses" become real when a specific invoice hits a specific account.

The moment of truth: simultaneous delivery, deemed delivery, and release instructions

Modern closings are usually virtual. Documents are circulated in PDF, signatures are collected via signature packets, and originals are handled later if needed. The key legal trick that makes virtual closings work is the concept of *deemed delivery*: documents are treated as delivered (and effective) when counsel releases them, often at the moment funds are released.

This is where funds flow and legal mechanics fuse. Counsel typically will not authorize wires until closing conditions are satisfied or waived. Banks typically will not send wires without an authorized instruction. Escrow agents typically will not release escrow without the right certificate or instruction. The process becomes a chain of "green lights," and the funds flow memo is the wiring map that everyone agrees is the correct destination once the lights turn green.

The real risk here is not fraud by insiders; it's misalignment by professionals. If counsel thinks "closing" means "documents released" while the bank thinks "closing" means "wires received," you can end up in an expensive limbo: documents effective but money not moved, or money moved but liens not released. A tight funds flow memo, paired with a clear release sequence, prevents that.

A square picture of a messy reality: money in, money out

The cleanest way to understand funds flow is to visualize it like plumbing: multiple sources feed a single junction (the paying agent/escrow), which then distributes funds to multiple uses. The diagram below is intentionally simplified but faithful to the closing-day logic.

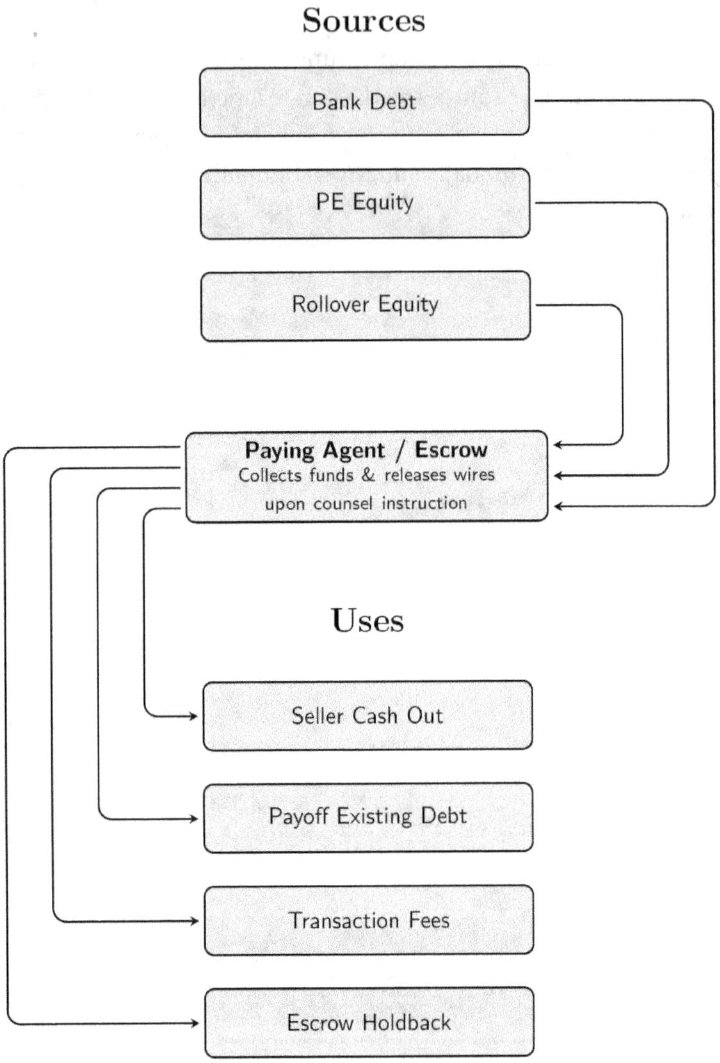

In the real world, each of those thick ribbons breaks into more ribbons. "Seller Cash Out" may be five owners in different jurisdictions. "Transaction Fees" may be ten wires, some domestic, some international, some to law firms that insist on separate trust accounts. "Payoff Existing Debt" may require separate wires to a lender and to a hedge counterparty for breakage. The diagram is still true: the work is consolidating complexity into a single, executable plan.

Who approves what: closing is a control problem disguised as a banking problem

If you want to predict whether a closing will feel smooth, don't ask who has the best lawyers. Ask who has the clearest approvals.

At minimum, you need:

- authorization to borrow (for bank debt),
- authorization to issue equity (for sponsor equity and rollover),
- authorization to disburse company funds (if the company is paying any expenses at closing),
- and a clear delegation of authority for who can approve wires and sign officer certificates.

Banks are allergic to ambiguity. If the CFO is authorized to instruct wires up to $5 million but the seller proceeds wire is $320 million, someone else must sign. If the signatory is on a plane, you have a problem that no amount of legal brilliance can solve. The most underrated closing workstream is simply ensuring the right humans will be reachable, sober, and documented.

Reverse termination fees and specific performance: the financial backbone beneath the pretty wires

Funds flow is the physical movement of money. Financing risk allocation determines whether the money *must* move.

If the buyer's remedy for failing to close is limited to a reverse termination fee (a negotiated payment to the seller), the buyer has effectively purchased an expensive option: it can choose not to close and pay the fee, even if the seller believes the buyer "should" have closed.

If the seller has the right to specific performance (a court order forcing the buyer to close) and that remedy is realistically enforceable, the buyer's freedom to walk is dramatically reduced.

This matters operationally because it changes how everyone behaves in the last two weeks. A buyer facing true specific performance tends to be obsessed with eliminating execution risk: payoff letters, lien releases, wire testing, signatures. A buyer that can cap its exposure sometimes tolerates more uncertainty, because the worst case is a fee, not ownership of a broken process.

Regardless of the remedy package, the funds flow memo becomes the shared object that makes "we are ready to close" a verifiable statement rather than a vibe.

Practical discipline: treat funds flow like a reconciliation, not a draft

The best closings treat funds flow like month-end close on steroids:

- lock the list of recipients,
- validate wiring instructions via callback procedures (never trust emailed account numbers without verification),
- reconcile totals to the purchase agreement math (including estimated adjustments and escrow),
- ensure payoffs match executed payoff letters,
- and run a timing plan that respects bank cut-off times.

The worst closings treat funds flow like a living document right up until release. That's how you get phantom wires, duplicate wires, and the dreaded post-close email: "We believe the amount sent to Lender B may have been short. Please advise." Nothing makes a newly closed deal feel less "closed" than realizing you might still owe a lender money.

The strangely comforting truth: precision is what makes closing feel safe

There's a reason experienced deal teams get almost superstitious about funds flow. Money movement is the one part of the process that is both utterly technical and deeply emotional. Sellers are saying goodbye to their company. Buyers are taking on risk. Lenders are betting on repayment. Everyone's heart rate is up, and the numbers have to be right anyway.

Funds flow is the antidote to that adrenaline. It replaces "we're paying $425 million" with a set of exact instructions that a bank can execute and auditors can trace. It's where the deal stops being a story and becomes a ledger. And on closing day, the ledger wins.

5.4 The Permission Slips: Regulatory and Third-Party Consents

Some deals fail for dramatic reasons: fraud, collapse, cold feet. Most delays happen for a quieter reason: someone who isn't in your conference room gets a veto. The government. A landlord. A customer's procurement team. A bank group that doesn't care how compelling your synergy deck is. These outsiders don't read your timeline; they impose their own. The transaction may be fully negotiated, fully signed, fully funded on paper—and still stuck behind a locked door until the right stamp arrives.

Regulatory approvals and third-party consents are called "conditions" in the contract, but they behave like *dependencies* in software: if you don't clear them, the program doesn't run. And unlike internal workstreams, you can't solve them with more late-night emails. You solve them by understanding what the gatekeeper actually wants, what it fears, and what process it must follow to protect itself.

The first mindset shift: this isn't about persuasion, it's about process

People often approach approvals as if they are miniature sales pitches: "If we just explain it well, they'll say yes." That works sometimes with business counterparties, almost never with regulators, and only occasionally with landlords. Most gatekeepers are not optimizing for your deal; they are optimizing for their own risk and compliance obligations.

A regulator's incentives are different from yours. It is designed to avoid being embarrassed later, not to maximize economic efficiency today. A customer's legal team doesn't get a bonus for helping you close; it gets blamed if it consents and then the service deteriorates. A landlord has a simple question: "Will the rent still get paid, and do I want this new tenant?" If the answer is "probably," they may still say "prove it."

So the craft here is not charisma. It's preparation: mapping each consent, determining whether it is truly required, sequencing outreach, and building a file that makes "yes" feel like the safest option.

HSR in plain English: the law that tells you to wait

In the United States, the best-known antitrust "permission slip" is the Hart-Scott-Rodino (HSR) process. In broad strokes, certain deals above size thresholds require the buyer and seller to file notifications with the FTC and DOJ, pay a filing fee, and then *wait* for a statutory period before closing. That waiting period is not a suggestion; it's a pre-close standstill. If you close early, you are not "bold." You are in violation.

The most important practical point about HSR is that it transforms time into a contractual risk. You can draft a purchase agreement in two weeks, but you cannot draft your way around a waiting period. Even in a clean deal, the clock has to run. In a complicated deal, the initial waiting period can be the beginning of the story, not the end.

The waiting period is the easy part; the "Second Request" is the plot twist

The phrase "Second Request" sounds like a follow-up email. It's not. It is the formal mechanism that extends the review and turns a manageable filing exercise into a major operational event. A Second Request typically demands a large volume of documents, data, and often detailed information about competition, customers, pricing, and strategy. It can dominate the interim period and consume management attention in a way that feels wildly disproportionate to the deal's press release.

This is where antitrust becomes a deal mechanics issue, not just a legal one. The interim operating covenants and information-access covenants suddenly matter more because the buyer needs the target's cooperation to respond. Meanwhile the business still has to hit numbers, retain talent, and avoid doing anything that looks like premature integration (more on that below). A Second Request can turn the "path to closing" into a tightrope: you have to move quickly, but not touch anything.

Who "gets" the deal: why two agencies receive filings but one drives the review

Even though filings go to both the FTC and DOJ, typically one agency takes the lead in substantive review. You don't get to choose. The agencies decide based on factors like industry familiarity and internal workload. That handoff can feel opaque from the outside, but the operational implication is simple: once you know who is leading, you tailor your engagement style accordingly and avoid trying to "forum shop" in a way that irritates everyone.

In well-run processes, the parties align early on an antitrust strategy: what markets are implicated, what the likely concerns are, and what remedies (if any) might be required. The worst time to discover you need a divestiture is when your outside date is approaching and the buyer's lenders are getting nervous.

The invisible antitrust constraint: "gun-jumping" and the taboo against acting like you've already closed

While HSR is pending, the parties are expected to remain independent competitors (if they are competitors) and independent businesses (even if they are not). "Gun-jumping" is the umbrella term for unlawful pre-closing coordination or premature transfer of control. You can share information to diligence and plan, but you can't start operating as one company.

This matters because the interim period is full of temptations:

- sharing forward-looking pricing plans,
- coordinating bids for the same customers,
- telling sales teams to "stay out of each other's lanes,"
- letting the buyer approve ordinary business decisions beyond what the signed covenants require.

The irony is that, in a well-intentioned attempt to protect the asset, buyers sometimes ask for operational control that creates antitrust risk. The cure is disciplined governance: clean teams for sensitive information, clear boundaries between consent rights (allowed) and management control (not allowed), and a habit of documenting that the seller is still running the business.

Remedies and "hell or high water": when approvals become an obligation, not a hope

Contracts don't just say "we'll seek approvals." They often say how hard each party must work to get them. The phrases range from "reasonable efforts" to "commercially reasonable efforts" to "reasonable best efforts" to "best efforts." The words can sound like synonyms;

they are often treated as bargaining chips that reflect how much pain a party is committing to endure.

In antitrust-heavy deals, the real fight is usually about remedies: divestitures, behavioral commitments, supply agreements, firewalls, or other concessions demanded to get clearance. A "hell or high water" clause is the extreme version, where the buyer commits to do whatever is necessary (up to a specified cap or in full) to obtain approval. Sellers love this because it pushes regulatory risk onto the buyer. Buyers fear it because it can force them to buy the company *and* give up the strategic benefits that justified the price.

The practical lesson is that regulatory risk isn't just "will we get approval?" It's "what will approval cost, who pays, and who decides whether the cost is too high?"

CFIUS: the national security permission slip that doesn't care about your market share

Antitrust asks whether a deal harms competition. CFIUS (the Committee on Foreign Investment in the United States) asks a different, more primal question: does this transaction create national security risk? If so, it can impose mitigation measures, recommend the President block the deal, or unwind it after the fact in certain circumstances.

CFIUS can be relevant not only for obviously defense-adjacent companies. Sensitive data, critical infrastructure, certain technologies, and proximity to sensitive facilities can all matter depending on the fact pattern. Some filings are mandatory in specified scenarios; many are voluntary but pursued because clearance provides meaningful protection against later review.

Operationally, CFIUS is its own calendar and its own document burden. It also tends to intersect with the deal's communications strategy. The wrong rumor at the wrong time can escalate political attention, which is never a fun variable to add to a closing timeline.

Industry regulators: the deal may be legal, but the license may not transfer

Beyond antitrust and CFIUS, sector regulators can quietly become the critical path. Banking deals can require approvals from federal and state regulators. Communications deals can involve FCC consent for license transfers. Healthcare can involve provider enrollment, certificate-of-need regimes, or state-level notice and consent rules. Energy and utilities bring their own alphabet soup.

These regulators often care less about price and more about stability: capital adequacy, managerial fitness, continuity of service, compliance history, community impact. That can be jarring for deal teams trained to speak in EBITDA. Sometimes the most persuasive "synergy" story to a regulator is not "we'll cut costs," but "we'll invest, maintain service levels, and strengthen compliance."

Also, these regimes can require *post*-closing notifications or filings. That's not a footnote. It affects the closing checklist, the integration plan, and the risk of running afoul of conditions that survive closing.

Third-party consents: the contracts that behave like landmines

Now for the approvals that feel less grand and more personal: third-party consents. These are required because many contracts contain anti-assignment provisions or specific change-of-control clauses. The simplest version says, "You may not assign this agreement without our consent." A more deal-relevant version says, "If there is a change of control of your company, we can terminate unless we consent."

The business impact can be wildly disproportionate to the paper. One customer contract might represent 18% of revenue. One lease might cover the only facility with specialized equipment. One software license might be embedded in every workflow. If those contracts don't survive closing, the buyer may not be getting what it priced.

Third-party consents are also where leverage shows up wearing a polite smile. Counterparties know you need them. Even if the contract says consent can't be "unreasonably withheld," they can slow-walk, ask questions, and create enough friction to extract concessions. The negotiation might not look like a shakedown; it will look like "due diligence." But the economic effect can be the same.

The three categories of consents (and why mixing them up is expensive)

Not every "consent" should be handled the same way. A disciplined approach separates them into categories:

Category 1: Must-have consents. If you don't get them, the deal can't close (either because they are conditions in the purchase agreement or because losing the contract would constitute a material issue). These deserve early outreach, senior attention, and tight tracking.

Category 2: Nice-to-have consents. The contract might technically require consent, but the counterparty is low-risk, low-value, or historically cooperative. These can be bundled, handled later, or addressed via alternative structures (where available) with counsel.

Category 3: "Consent not required" positions. Sometimes the best strategy is to conclude, with legal support, that no consent is required (for example, because there is no assignment in a stock sale, or because the clause is poorly drafted). This is a real strategy, not bravado, but it must be chosen carefully. Getting it wrong can trigger termination rights after closing, which is the worst time to discover you misread a contract.

The mistake is treating every contract as Category 1. That floods the counterparty universe with "We're being acquired" emails, increases the chance of leaks, and creates self-inflicted delays. The opposite mistake is assuming everything is Category 3 and discovering at the eleventh hour that the landlord has a veto and a lawyer who enjoys using it.

Change-of-control outreach: the art of asking without spooking

Asking a customer for consent is basically telling them, "Your supplier is about to change owners." They hear: "Service might get worse." Asking a vendor for consent can sound like, "We're bigger now, and we might squeeze you." Asking a landlord for consent can sound like, "We're about to change the credit behind this lease."

So the outreach has to be carefully framed. The most effective messages are short, concrete, and protective of the counterparty's interests:

- continuity of service,
- continuity of account management,
- no change to pricing or terms as part of the consent request (unless you want a negotiation),
- and a clear statement of what, exactly, they are being asked to consent to.

The biggest unforced error is letting the first outreach be a 12-page legal document with signature blocks and no human context. That's how you get routed to the slowest possible internal queue.

The "most favored nation" trap: consents that accidentally reopen your whole commercial model

Some counterparties treat consents as an invitation to renegotiate. Sometimes they are explicit about it: "We'll consent if you extend the term and reduce price." Sometimes they are subtle: they ask for updated pricing schedules, audit rights, security addenda, expanded SLAs, or insurance certificates that effectively change the deal.

The most dangerous version is when one customer requests better terms, and granting them triggers a "most favored nation" clause elsewhere. Suddenly a small consent concession becomes a cascading repricing event. This is why consent workstreams need someone who understands not only the legal text but the commercial interdependencies.

A good practice is to run a "concession control" process: any term change offered to secure consent is reviewed for downstream impacts (MFNs, margin effects, operational obligations) and for consistency with the purchase agreement (which may restrict amendments to material contracts without buyer consent).

Timing risk is not just delay; it's optionality decay

Delays aren't neutral. The longer it takes to obtain approvals and consents, the more the business operates under interim constraints, the more uncertainty leaks into employee morale, and the more chances there are for performance to drift. That drift can feed into other deal mechanics: covenant compliance, bring-down standards, and even the parties' appetite to litigate or re-trade.

This is why sophisticated teams treat approvals and consents as a *critical path* exercise with real management attention. It's not merely a legal checklist. It's a value preservation effort. A deal stuck in limbo is a deal where your best employees answer recruiters' calls, where customers hedge their bets, and where competitors start selling "certainty" against you.

A practical consent map: what to track so you don't learn about a veto on closing week

A tight tracking document for consents is not complicated, but it has to be specific. At minimum, it should capture for each consent:

- the contract name and counterparty,
- the clause that triggers consent (assignment, change of control, both),
- whether the transaction structure triggers it (asset sale vs. stock sale vs. merger),

- the business owner responsible for the relationship,
- the outreach plan and timing (who asks, how, and when),
- the counterparty's response status,
- and any concessions requested or implied.

The value of this map is not administrative neatness. It's early warning. A counterparty that doesn't respond for two weeks is not "busy"; it's a risk. A counterparty that asks for a call with its CFO is not being friendly; it's escalating. The map makes those signals visible early enough to manage.

When you can't get consent: workarounds, risk decisions, and grown-up conversations

Sometimes you simply can't get a consent in time. The counterparty stalls. The regulator extends the process. The landlord wants a guarantee the buyer won't give. At that point, the deal team has to stop hoping and start choosing.

The menu of responses is finite:

- *Wait*: extend the outside date, accept the business risk of delay, and keep pushing.
- *Restructure*: change the transaction structure to avoid assignment in certain jurisdictions or contracts (where legally viable).
- *Carve out*: exclude an asset, contract, or business line from the deal and address it separately.
- *Condition/waive*: treat the consent as a closing condition (or remove it), with price adjustments, escrows, or indemnities to allocate the risk.

None of these is painless. The point is that consent problems are not solved by pretending they aren't real. They're solved by allocating the consequences explicitly: who bears the risk if the contract terminates, who funds any replacement arrangement, and what happens if the loss triggers broader issues (like financial covenant breaches or customer churn).

The paradox: the best approvals work is quiet, and that's the point

A well-managed approvals and consents process is almost invisible to outsiders. Regulators get clean submissions and predictable responses. Customers get calm messages and continuity assurances. Landlords get clear financials and straightforward guarantee terms. The deal team doesn't brag about it because there's nothing to brag about; the plane simply lands.

When approvals go badly, everyone hears about them, because the sound they make is delay. In M&A, delay is rarely just lost time. It's lost certainty. And lost certainty is the one cost that spreads everywhere: into employee decisions, customer behavior, lender confidence, and the parties' willingness to keep acting like partners instead of litigants.

5.5 Crossing the Finish Line: The Closing Process

Closing used to be a ritual: people in suits around a table, pens that cost too much, an overconfident hand shake. Today it's an inbox flood and a conference call where someone says, "We are live," and everyone tries not to breathe until the wires hit. The modern closing has the vibe of a spacecraft launch: hundreds of small checks, each individually boring, collectively decisive.

The strange thing is that closing day rarely rewards brilliance. It rewards *mechanics*. The deal doesn't close because the parties feel aligned. It closes because every required document is in the right form, every condition is checked or waived, and the people controlling the money have unambiguous instructions to release it. When that machinery works, closing feels effortless. When it doesn't, you learn exactly how fragile "ready to close" can be.

Signing vs. closing: the two most commonly confused verbs in M&A

A deal can be *signed* without being *closed*. Signing is when the parties commit by executing the definitive agreement. Closing is when ownership actually changes hands and the consideration is actually paid. In a simultaneous sign-and-close, those moments are basically the same breath. In a sign-then-close deal, they can be weeks or months apart, with the business operating in a constrained interim state.

The confusion matters because people act differently based on which verb they think they're living in. Employees hear "we signed" and assume it's done. Customers hear "we're closing soon" and wonder if service is about to change. Banks hear "we're closing today" and ask, quietly and relentlessly, "Is that *today* before cut-off, and do you have release instructions?" The only safe rule is to treat "closing" as a technical event, not a mood.

The closing checklist: the deal's nervous system

Every real closing is driven by a closing checklist, usually managed by deal counsel. It is a list of required deliverables, organized by party and by document type, with status notes that start optimistic and end terrifying. The checklist is not a bureaucratic artifact; it is the closest thing to an operating system the deal has.

A good checklist does two things at once:

- It forces you to identify every required piece of paper early enough to fix it.
- It creates a shared reality: everyone can see what is missing, who owns it, and what "done" means.

The reason experienced teams love checklists is that closings fail in mundane ways. Not because the buyer suddenly hates the company, but because someone's certificate wasn't dated, a board consent didn't authorize a specific act, or a payoff letter didn't include the lien release language the buyer's lenders require. The checklist is how you prevent a $400 million process from being held hostage by a missing initial.

Virtual closing: the email avalanche that replaced the boardroom

In a virtual closing, the closing room is a shared drive, a PDF binder, and a sequence of emails that include the word "EXECUTED" in the subject line. That sounds casual until you watch it happen. A coordinated virtual closing is intensely structured: counsel controls the flow, documents are compiled in designated folders, and signature pages are marshaled like aircraft on a runway.

The operational trick is that everyone must agree what counts as "delivery." PDFs are generally acceptable if the agreement says so (and most modern agreements do). But third parties may still require originals: a wet-ink signature for a particular filing, a notarized document, or an original stock power. A virtual closing is not a reason to be sloppy about formalities; it's a reason to track them more carefully, because your intuition isn't being guided by the physical act of handing paper across a table.

Signature packets: why you sign pages that aren't attached to anything (yet)

One of the most unintuitive closing mechanics is the signature packet. Parties sign signature pages in advance, and those pages are held "in escrow" by counsel. Later, once everyone agrees that closing conditions are satisfied, counsel attaches the signature pages to final versions of the documents and releases them.

This sounds like a loophole until you realize what it accomplishes: it decouples the human bottleneck from the legal moment. If you had to get every executive on the phone at the exact moment the last condition cleared, closings would be impossible. Signature packets allow people to sign during daylight hours, from stable internet, with time to fix errors, while still preserving a controlled release moment.

Signature packets also create a quiet discipline: if someone signs early, they are implicitly authorizing counsel to use that signature page only when the agreed conditions are met. That's why counsel are so careful about version control. A signature page attached to the wrong version of a document is not a cute mistake; it can be a contractual disaster.

The bring-down moment: the last time the seller says, ``Yes, still true"

Right before closing, the parties "bring down" key facts. This is the contractual equivalent of a pilot's final pre-flight check. The seller typically delivers officer certificates confirming that representations are true to the required standard and that covenants have been performed as required. The buyer may deliver similar certificates for its own reps and covenants.

These certificates matter because they concentrate risk. A representation in the agreement can feel abstract; an officer certificate feels personal. It is a named individual signing a statement that will live forever in the closing binder. That tends to sharpen internal diligence: finance teams make sure numbers reconcile, legal teams confirm no new litigation has appeared, and executives ask uncomfortable questions they should have asked earlier.

The timing is also practical. Certain supporting documents are time-sensitive: good standing certificates, secretary's certificates, payoff letters, insurance certificates. You want them fresh enough to be reliable, but not so fresh that you can't replace them if something is wrong. Many teams aim for a "T-minus" rhythm: order and draft early, finalize close to closing, and leave a buffer for the universe's sense of humor.

Funds release: closing is often a sequence, not a single click

Even when people talk about closing as a single moment, it often occurs as a sequence of authorizations:

- Counsel confirms conditions are satisfied or waived.
- Counsel authorizes release of signature pages (deemed delivery).
- The bank receives final wire instructions and any required sign-offs.
- The paying agent or escrow agent confirms receipt of incoming funds.
- Outgoing disbursements are released in the agreed order.

Order matters because money and liens are entangled. If the buyer is paying off existing debt, it wants lien releases in return. Lenders often commit to release liens upon receipt of funds, but they still need to *receive* the funds. Meanwhile the seller wants its proceeds. A paying agent helps, but the underlying reality remains: everyone is trading something valuable, and nobody wants to go first without protection.

This is why closings are governed by "release instructions," often in writing from counsel. Banks and escrow agents want a clear directive: what they are authorized to do, under what conditions, and who bears the risk if something is contested later. On closing day, ambiguity is the enemy; it makes cautious institutions become immovable objects.

The closing call: twenty people, one sentence that matters

The closing call is usually a short conference call with a long attendance list: buyer counsel, seller counsel, sometimes bankers, sometimes the escrow agent, sometimes the lender's counsel. The call is not for negotiation. It is for synchronization.

A well-run closing call has a clear final checklist review: any open items, any deliveries pending, confirmation that funds are in position, confirmation of wire release timing, and then the moment where counsel says the words that flip the deal from "planned" to "effective." There is often an almost comical anticlimax: a few confirmations, a pause, and then, "Okay, we're closed." If you hear cheering on the call, it usually means there was real fear five minutes earlier.

One reason the closing call matters is psychological. Deals generate anxiety, and anxiety generates divergent assumptions. The call forces everyone to agree on the same reality at the same time. It is a small ritual, but it prevents a lot of post-close emails that start with "We thought..."

Filing and recording: the deal is closed, but the world still needs to be told

Depending on structure, closing may require filings or recordings to perfect the transaction. In a merger, there may be a certificate filed with a Secretary of State. If debt is being paid off, there may be UCC terminations to release security interests. If new debt is being put on, there may be new UCC filings. Some industries require post-consummation notices to regulators.

These tasks can feel like administrative loose ends, but they're not. Filings are how you make the closing "real" to third parties. A lender's lien doesn't disappear because you feel like you paid it; it disappears because the release is properly delivered and, when required, properly filed. A corporate merger doesn't exist because your board approved it; it exists because the state accepted the filing.

The operational lesson is to treat post-close filings like part of closing, not part of integration. Assign owners, track deadlines, and confirm completion. The deal may be economically done, but if legal perfection isn't handled, you can create problems that surface months later when someone tries to refinance, sell an asset, or prove clean title.

Closing deliveries that everyone forgets until they're missing

A surprising number of closing scrambles are caused by documents that are obvious only in hindsight:

- *Payoff authorization letters* from the borrower instructing its lender to accept payoff and release liens.
- *Resignation letters* for directors or officers changing at closing, especially in private company deals.
- *Stock certificates and stock powers* (or evidence of their cancellation) if equity is transferring.
- *FIRPTA certificates* in deals involving U.S. real property interests (a classic "please don't bring this up on closing morning" item).
- *Updated schedules* reflecting changes since signing, where permitted, and the internal debate about whether they are permitted.

The reason these get forgotten is that they live at the seams between disciplines: corporate governance, tax, treasury, employment, and operations. No single person "owns" them by instinct. That's why the checklist exists, and why good deal teams obsess over it.

The human risk: closings are where fatigue becomes a financial variable

By the time you reach closing, everyone is tired. Tired people miss details, and details at closing have dollar signs attached. The most dangerous hours are often the late ones, when someone suggests "we can fix that after close." Sometimes you can. Often you can't without reopening the deal or creating an immediate breach.

The antidote is not heroics; it's pacing. Build buffers. Avoid "closing at 5:00pm" if your payoff letters are only good through 5:00pm. Avoid last-minute changes to funds flow. Avoid introducing a new form of escrow instruction the morning of closing. Each novelty adds cognitive load, and cognitive load is where mistakes hide.

Also: be kind to the people doing the work. Closings fail when the team stops communicating cleanly. A CFO snapping at counsel doesn't make documents arrive faster; it makes people more cautious, and caution slows everything.

The "deal is closed" email: the smallest message with the biggest consequences

At some point after the releases and wires, someone sends the email: "We are closed." It's a short sentence that does a lot. It tells the integration team they can finally act. It tells the seller they can exhale. It tells lenders they can book the loan. It tells auditors where the line is between "before" and "after."

It also creates a timestamp that matters. For interest calculations, for working capital mechanics, for per diem on payoffs, for determining whether a condition was satisfied "as of closing," for regulatory notifications. That's why closing emails often include the exact time and time zone, even when it feels pedantic. In deal mechanics, pedantry is just accuracy wearing a better outfit.

The best closings feel almost dull. The documents line up. The wires move. The lien releases are authorized. The right people are reachable. The final email lands and nobody argues about what it means. If that sounds unromantic, good. Romance is how you sign. Precision is how you close.

Chapter 6

Post-Closing and Special Situations

The champagne is dry, the signatures are ink-stained, and the wire transfer has cleared. Is the deal done? Not quite. In this final chapter, we navigate the critical 'morning after' phase-where the estimated price meets financial reality, and two companies attempt to become one. We also venture off the beaten path to explore the high-wire acts of corporate carve-outs and distressed sales, where standard rules don't apply and speed often trumps perfection.

6.1 The Price Tag Was Just an Estimate: True-Ups and Disputes

At closing, everyone wants the same thing: wire the money, swap signatures, take the awkward group photo, and go home. The problem is that the most fought-over number in the entire deal often can't be known with certainty at that moment. Not because anyone is hiding the ball (though sometimes they are), but because the purchase price in many deals is built on a moving target: the company's day-to-day operating balance sheet, measured at a specific instant in time, using accounting judgments that reasonable people can disagree about.

That's the emotional whiplash of post-closing adjustments. You just paid "the price," yet the contract quietly insists it was only a first draft. The deal has a tab open.

Estimated vs. Actual: the closing statement is a snapshot, not a sculpture

The closing mechanics usually force a compromise between precision and momentum. The buyer needs to pay a number on closing day. The seller needs to deliver the company cleanly on closing day. But the accounting information needed to calculate certain elements of that number—especially Net Working Capital (NWC)—is typically not final on closing day. Month-end closes aren't finished, invoices are still arriving, payroll accruals are still being computed, inventory counts are still being reconciled, and someone is still arguing about whether a particular receivable is collectible or just *optimistic fiction*.

So the parties do what deal people always do when reality won't fit neatly into a deadline: they agree to estimate now and reconcile later.

The typical structure is a two-step dance:

1. An *Estimated Closing Statement* is prepared right before closing, producing an estimated purchase price (or estimated adjustment to a base price).
2. After closing, the buyer prepares a *Final Closing Statement* based on actual numbers as of the closing moment, then the parties true-up—a post-closing payment either from seller to buyer or buyer to seller.

That sounds tidy. In practice, it's one of the most common ways a relationship that was barely polite during negotiations becomes actively hostile after the wires clear.

The spark is simple: the estimate gets you to closing; the final numbers decide who owes whom real money.

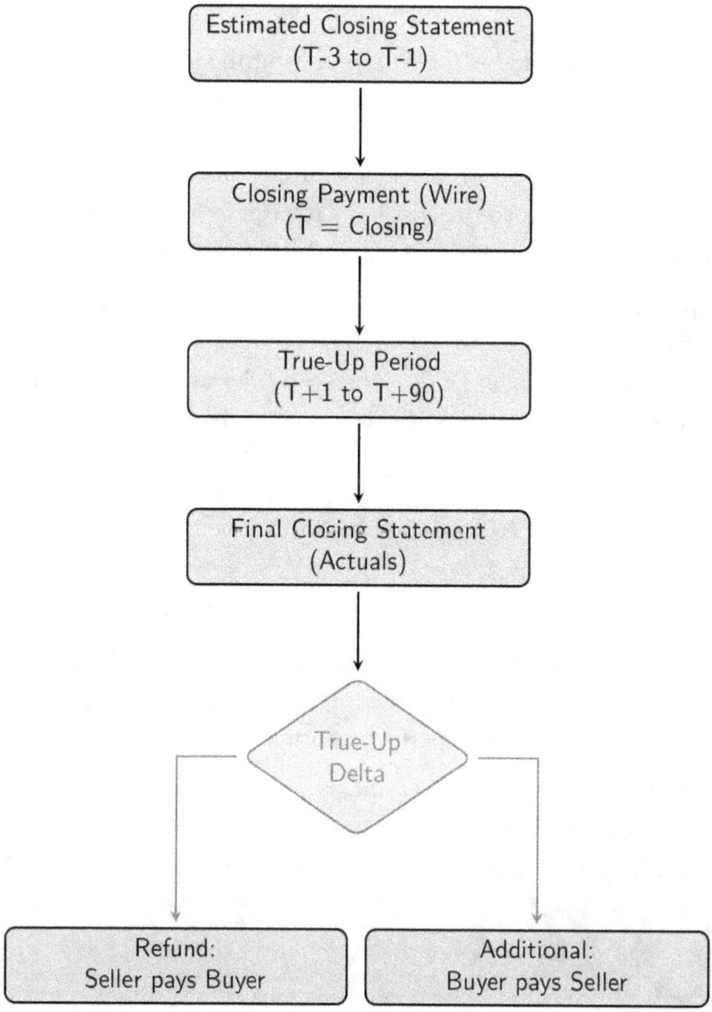

Estimated at closing, corrected after books are finalized. The highlighted delta can move in either direction.

Why working capital is the usual culprit

Net Working Capital is the short-hand for the operating "float" inside the business: the current assets and current liabilities that keep the machine running day to day. In plain English, it's the money tied up in things like receivables and inventory, net of things like payables and accrued expenses. It's not cash itself; it's the set of accounts that explain why cash will be higher or lower next month.

Buyers care about NWC because it's the difference between buying a business that can breathe on Day 1 and buying a business that immediately needs an IV drip. If receivables are thin, inventory is understocked, or payables have been aggressively paid down before closing, the buyer can inherit a company that technically "closed" but can't fund its own operating cycle without the buyer injecting cash right away.

Sellers care about NWC for the opposite reason: it's easy for a buyer to weaponize the calculation after closing. If the definition is fuzzy, the buyer can argue that "working capital was short," and quietly claw back purchase price in the form of a true-up, even if the business was handed over in perfectly normal operating condition.

That's the core tension: NWC adjustments are meant to preserve the economic deal, but they can become a post-closing battlefield where accounting choices substitute for negotiating leverage.

The ``peg" is the deal's thermostat---and everyone fights over the temperature

Most working capital true-ups compare actual closing NWC to a target called the *peg*. The peg is not a moral statement about what "good" working capital looks like. It's a negotiated baseline intended to reflect the level of working capital required to run the business in the ordinary course.

If actual NWC at closing is above the peg, the buyer is receiving extra operating value and typically owes the seller additional purchase price. If actual NWC is below the peg, the buyer is receiving less operating liquidity than promised and the seller owes money back.

This sounds mathematical until you remember what "ordinary course" means in real life. Many businesses don't have a single "normal" working capital level. They have seasons. They have one-off customer programs. They have lumpy inventory buys. They have customers who pay early when they feel like it and late when they can get away with it. A peg can be crafted thoughtfully (based on historical averages, seasonality, and the quality of earnings work) or it can be negotiated like a hostage exchange ("split the difference and stop calling me").

A peg that's wrong doesn't just create a pricing mismatch. It creates a predictable post-close dispute, because it forces the true-up to "correct" something the parties never agreed on in the first place.

The hidden lever: definitions beat spreadsheets

A working capital schedule can look precise while being fundamentally undefined. Two teams can start with the same trial balance and produce different NWC numbers without anyone committing fraud. The divergence comes from classification, methodology, and policy choices that aren't visible in a one-line "NWC" figure.

The most common fault lines are not exotic:

- *What counts as current?* A liability due in 14 months isn't current under many accounting frameworks, but it might still drain near-term liquidity depending on covenants, payment patterns, or looming disputes.
- *What gets excluded?* Transaction bonuses, deal fees, and restructuring reserves often get carved out, but the carve-outs must be explicit or they become a creativity contest.
- *What happens with deferred revenue?* In some businesses, deferred revenue is a core working-capital item; in others, it be-

haves like a quasi-debt (future service obligation with real cost).

- *How are reserves handled?* Bad debt reserves, inventory obsolescence, warranty accruals: these are judgment calls disguised as decimals.

Here's the uncomfortable truth: the "right" working capital number is less important than the *agreed* working capital number. A true-up mechanism is not an audit of the seller's historical accounting. It's a contractually defined calculation that tries to translate economics into a repeatable formula. When the formula isn't repeatable, the true-up becomes a proxy war about the whole deal.

GAAP vs. consistency: the clause that looks boring and costs millions

If there is a single sentence that should make a deal team sit up straighter, it's the one describing the accounting standard for the closing statement. Parties often try to be "reasonable" by requiring that the closing statement be prepared in accordance with GAAP. That seems sensible until you realize GAAP is not one thing. It's a universe of choices, estimates, and elections, and a company's historical financial statements may be GAAP *and* still reflect long-standing practices that are imperfect, conservative, aggressive, or just idiosyncratic.

So deal lawyers add another concept: consistency with past practice. Now you have two masters: GAAP and consistency. When they align, life is easy. When they conflict, the dispute writes itself.

Consider the incentives:

- A buyer who suspects the seller historically used aggressive accounting may push for strict GAAP to "clean up" the numbers at close—which can lower working capital and trigger a refund to the buyer.

- A seller who fears the buyer will "re-underwrite" the deal through accounting changes will push for consistency with the seller's historical methods—which can preserve the seller's expected economics.

The drafting solution is not philosophical; it's hierarchical. If GAAP and consistency both appear, the agreement should say which prevails, and ideally for which line items. Otherwise you are asking an independent accountant (or a judge) to solve a problem the contract could have solved with one additional sentence.

Estimated closing statements: the art of being wrong in a defensible way

The Estimated Closing Statement is often prepared by the seller, sometimes with input from the buyer, under crushing time pressure. It's not unusual for it to be assembled from a combination of partial month results, management estimates, and a good-faith attempt to map what's happening in the business onto a spreadsheet that was negotiated weeks earlier.

The goal is not to be perfect; the goal is to be *credible*. A credible estimate does three things:

1. Uses the agreed methodology, even when it's inconvenient.
2. Shows its work (support schedules, roll-forwards, logic for accruals).
3. Avoids "heroic" assumptions that only benefit one side.

An estimate that is strategically optimistic doesn't just increase the odds of paying money back later. It also poisons the post-close process, because it signals that the other side can't trust your numbers. Once trust is gone, every invoice becomes a deposition exhibit.

Final closing statements: the buyer has the pen, and that changes the temperature

After closing, the buyer typically prepares the Final Closing Statement. That makes practical sense: the buyer now owns the books and controls the accounting team. It also creates a predictable anxiety for sellers: the other side has the pen and a financial incentive to press it hard.

A well-run buyer treats the final statement like a regulated product: documented assumptions, consistent policies, and a clear mapping from the agreement's definitions to the company's general ledger. A poorly run buyer treats it like a negotiation do-over, which is how you end up paying outside advisors to argue about whether prepaid insurance is "operating" or "non-operating" working capital.

The post-close adjustment is supposed to be a narrow, mechanical process. The moment it becomes an excuse to revisit the purchase price, it stops being a closing statement and starts being litigation in spreadsheet form.

The calendar matters: the true-up period is a pressure cooker

True-ups usually have a defined timeline: the buyer delivers the final statement within a certain number of days, the seller has a review period, disputes are noticed, and unresolved items go to an independent accountant. Many deals use something like 60 to 120 days for the buyer to prepare the statement and a shorter window for the seller to object.

Those are not just procedural dates. They determine leverage.

A buyer who drags its feet can force the seller into a rushed review, especially if the seller's deal team has disbanded and the CFO has moved on to the next job. A seller who delays objections can miss notice deadlines and lose the right to fight. A contract that looks "standard" on timing can become outcome-determinative if one side is organized and the other is exhausted.

The practical lesson is almost annoyingly simple: treat the true-up like it's part of closing, not an afterthought. Keep the models, keep the workpapers, keep the people on standby, and assume that your counterparty's memory will be selective.

The dispute letter: where tone becomes a legal strategy

When the seller objects to the Final Closing Statement, it typically sends a dispute notice identifying specific items and the basis for disagreement. That letter is not a venting session. It is the first document a neutral decision-maker will read if the dispute escalates.

The best dispute letters are almost boring:

- They cite the agreement's definitions and the specific line item.
- They show the buyer's number, the seller's number, and the delta.
- They attach support.
- They avoid accusations unless the agreement actually supports them.

It's tempting to write, "Buyer is manipulating the calculation." It's usually more effective to write, "Buyer's treatment of accrued rebates is inconsistent with Schedule X and the illustrative example; applying the agreed methodology yields $Y."

The independent accountant is not there to referee your feelings. It is there to apply a contract to a set of numbers.

Independent accountants: arbitration in a green visor

Most purchase agreements do not send true-up disputes to court. They send them to an independent accountant (often a major accounting firm) for binding resolution. This is sometimes described as "arbitration," but it behaves differently than courtroom litigation. The independent accountant's authority is usually limited to deciding disputed items within a defined scope: the calculation of the adjustment under the agreement.

That scope limitation is not a footnote. It's the dividing line between:

- *Accounting disputes*: classification, application of agreed policies, mathematical errors.
- *Legal disputes*: what the contract means, whether an item is excluded at all, whether fraud or covenant breaches occurred.

When that dividing line is unclear, parties fight about the fight: "Is this an accounting issue or a legal interpretation?" You can burn months and seven figures arguing about jurisdiction before anyone touches the underlying numbers.

A tight process avoids that by defining:

1. The independent accountant's role (expert determination vs. arbitration).
2. The issues it can decide (and cannot decide).
3. The procedures (submissions, meetings, timelines, whether it can request information).
4. The standard (GAAP, consistency, hierarchy).

The punchline is that independent accountants are not cheap. But compared to a full-bore court fight, they are often faster and more predictable—especially when the contract gives them a clear lane.

What true-up disputes are *really* about

On paper, a working capital true-up is about current assets and current liabilities. In reality, true-up disputes often reveal three deeper anxieties.

First: buyer's remorse. A buyer that discovers operational problems post-close can look for a contractual mechanism to claw back value.

The true-up is attractive because it feels technical, and technical arguments can be made to sound inevitable.

Second: seller's fatigue. Sellers underestimate how much effort it takes to defend a closing balance sheet after the deal team has scattered. The true-up becomes a test of endurance: who is willing to keep paying advisors and management time?

Third: control of the narrative. The party with the books post-close can shape the story of what the numbers "mean." The other party must reconstruct that story from reports, data rooms, and whatever cooperation the agreement compels.

If you want to predict where a dispute will land, don't start with the spreadsheet. Start with incentives, bandwidth, and access to information.

A concrete example: how a normal business turns into a $2 million argument

Assume a deal with:

- Purchase price: $100 million (cash-free, debt-free structure)
- NWC peg: $12 million
- Estimated closing NWC: $12.5 million (seller's estimate)
- Final closing NWC: $10.5 million (buyer's calculation)

Under the typical mechanics, the buyer would claim a $2.0 million shortfall versus the peg and demand a $2.0 million payment from the seller (often from escrow, if available).

How do you get from $12.5 million estimated to $10.5 million final without anyone forging documents?

- The buyer reclassifies a large prepaid as non-operating.
- The buyer increases the bad debt reserve based on post-close collections (a classic point of contention).

- The buyer includes an accrual for customer rebates that the seller historically tracked off-book until invoiced.

Each of those can be argued as "reasonable." Each can also be argued as inconsistent with historical practice or the agreed example schedule. Multiply "reasonable" by a few line items and you have a multi-million-dollar delta.

This is why illustrative schedules matter. A peg is a target; the schedule is the map. Without the map, you're debating directions with someone who controls the GPS.

Drafting that prevents disputes: specificity is kindness

Preventing true-up disputes is not about writing longer definitions. It's about writing definitions that reduce interpretive room where interpretive room is expensive.

The best agreements tend to do a few unglamorous but powerful things:

- *Define working capital by line item.* Not just "current assets minus current liabilities," but a schedule of included and excluded accounts.
- *Include an illustrative calculation.* One example, prepared from a real historical period, using the exact format that will be used post-close.
- *Set an accounting hierarchy.* If it's GAAP, say GAAP. If it's consistency, say consistency. If it's "GAAP applied consistently with past practice," say which wins when they clash.
- *Address estimates.* Spell out how accruals are handled, whether post-close information can be used, and which cutoff procedures apply.

There is a reason sophisticated buyers and sellers obsess over the working capital schedule. It's not because they love accounting. It's because the schedule is a purchase price lever that activates after the champagne.

Process discipline: the fastest way to lose is to be sloppy

Even with good drafting, the process can still go sideways if the mechanics are treated casually.

For buyers:

- Lock the methodology early and communicate it to the accounting team.
- Preserve the closing-date data (AR aging, inventory reports, AP listings) as it existed at close. Rebuilding later is harder than it sounds.
- Prepare the final statement as if a neutral expert will read it—because one might.

For sellers:

- Keep your deal model and support schedules intact, with version control. "We can't find the file" is not a defense strategy.
- Use your review window aggressively. The clock is not your friend.
- Identify disputes with precision. Broad objections often die on procedural grounds.

A true-up is one of the few parts of M&A where operational competence and legal drafting collide head-on. The party that treats it like a project instead of a quarrel usually wins.

When the true-up becomes a proxy for indemnity

There is a recurring temptation in deal disputes: if the buyer feels cheated, it may try to shove the complaint into the true-up process because it's faster than an indemnification claim and often has a different burden of proof.

But the true-up is not designed to compensate for everything that can go wrong in a deal. It's designed to adjust for a specific set of balance sheet items at a specific time under a specific methodology. Trying to use it as a Swiss Army knife can backfire, especially if the agreement limits the independent accountant's authority to narrow accounting determinations.

A clean separation of remedies matters:

- *True-up* fixes the math of the closing balance sheet measurement.
- *Indemnification / reps & warranties* address breaches, misstatements, and other risk allocations.

When those boundaries blur, the dispute is no longer about working capital; it's about whether the buyer can re-price the deal after closing. That's where "technical" disagreements suddenly become existential.

Escrows and setoff: where the money sits changes how hard people fight

A true-up is also a collections problem. If the seller owes money, how does the buyer get paid? If the buyer owes money, how does the seller ensure it gets paid?

Many deals use an escrow to secure indemnification claims and sometimes true-up payments. If the true-up is payable from escrow, disputes can intensify because the buyer has an easier path to cash and the seller feels the money is being taken before the argument is fully heard. If there is no escrow and the seller has distributed proceeds, the buyer may face a real-world challenge collecting, which can make the buyer more aggressive on procedural rights and deadlines.

Setoff rights (the ability to net amounts owed under different parts of the agreement) are another accelerant. They can be efficient, but they can also turn a working capital fight into a multi-issue standoff: "We won't pay your earnout until you pay our true-up," or the reverse.

The mechanics of payment are not administrative. They are leverage disguised as wiring instructions.

A closing thought that's painfully practical

If you want the true-up to be boring—and you do—you have to decide what "boring" means before you close. Boring means:

- the definitions point to one answer more often than they point to a debate,
- the examples match the business as it actually operates,
- and the dispute process is narrow enough that it can't be hijacked into a re-trade.

The irony is that the most effective way to prevent a post-closing fight is to spend more time on the part everyone wants to rush. The closing statement looks like accounting. It's actually deal economics with a calculator. And calculators, unlike people, don't care how tired you are.

6.2 Day One and Beyond: Integration and TSAs

Closing day has a way of making everyone feel like the hard part is over. The signatures are in. The money moved. Someone's inbox fills with cheery "Congratulations!" emails written by people who will not be answering the calls at 2:00 a.m. when the payroll file fails to transmit.

Day 1 is when the deal stops being a document and starts being a company again. Customers still place orders. Employees still forget passwords. A regulator still wants a report in the same format as last quarter. Trucks still show up at the dock whether the ERP is ready or not. Ownership changes hands in one instant; operations don't.

That mismatch creates the central post-close operational question: if the seller's infrastructure was quietly doing a thousand jobs behind the curtain, what happens when the curtain is pulled down?

Day 1 readiness: the first morning you can't call it ``their problem"

"Day 1 readiness" sounds like consultant jargon until you translate it into plain language: can the business function tomorrow with a different owner?

The checklist is not glamorous, but it is unforgiving. Who issues customer invoices? Who collects cash? Who pays vendors? Who runs payroll? Who grants system access to a new hire? Who answers a cyber incident alert? Who files sales tax returns in seven states that all have their own opinions about what "on time" means?

A mature company can survive a lot of strategic confusion. It cannot survive not paying employees, not billing customers, or losing access to its own systems. Those failures don't feel like "integration risk." They feel like a business that is suddenly hemorrhaging trust.

This is why integration planning starts before closing even when the parties pretend they're too busy for it. The business doesn't care that the legal team is still redlining. It will demand the same services on Day 1 that it demanded the day before closing.

The TSA: renting the seller's back office like an apartment with furniture

A Transition Services Agreement (TSA) is the deal's most practical confession: the buyer doesn't yet have everything needed to run the target on its own.

The TSA is a contract where the seller agrees to provide specified services to the target (now owned by the buyer) for a defined period after closing. Think of it as renting the seller's operational scaffolding while the buyer builds its own. That scaffolding can include:

- IT hosting and helpdesk support
- ERP and accounting systems access
- Payroll processing and benefits administration
- Procurement, vendor management, and shared logistics
- Cybersecurity monitoring and incident response
- Facilities, real estate services, or even manufacturing support

A TSA is not a favor. It's not "support." It is a temporary supply agreement with consequences. The seller becomes a service provider to a company it no longer owns, while the buyer becomes dependent on a counterparty whose incentives have changed overnight.

And that's the tension humming beneath every TSA: the seller wants to unwind quickly and cheaply; the buyer wants continuity and time.

Why TSAs exist even when everyone hates them

TSAs persist because the alternative can be worse.

If the target was deeply embedded in the seller's platforms—shared HR systems, shared ERP, shared cybersecurity tools, shared vendor contracts—then "clean separation at close" is a fantasy unless someone spends a great deal of money and takes a great deal of risk before the deal even closes.

Sometimes that investment makes sense. Often, it doesn't. There are limits to how much stand-alone infrastructure you can build pre-close when you don't yet own the business, can't fully control its priorities, and can't freely migrate data without stepping into privacy, security, or contractual traps.

So the TSA becomes the bridge. Not because it is elegant, but because it is available.

The trick is remembering what bridges are for: getting off them. The buyer's goal is not to run the acquired business on the seller's systems indefinitely. The buyer's goal is to keep the lights on while building independent capability. A TSA that quietly becomes permanent is a slow-motion integration failure disguised as "business as usual."

The first TSA negotiation mistake: treating it like boilerplate

TSA templates are everywhere. That's part of the danger. The base document often looks standardized: term, fees, service schedules, governance, termination rights. The real deal is in the schedules—the service descriptions, service levels, pricing mechanics, and the assumptions buried in footnotes that no one reads because the lawyers are busy arguing about indemnity caps.

A TSA schedule that says "IT support" is not a service. It's a misunderstanding waiting to mature into an outage.

Services need to be described like you're writing an instruction manual for a stranger who doesn't care about your deal:

- What exactly will be done?
- By whom (named team, function, or third party)?
- During what hours, with what response times?
- Using what systems, with what access rights?
- What is explicitly out of scope?

Clarity feels tedious until the first incident ticket gets bounced between teams because "that's not in the TSA." Then clarity feels like oxygen.

Service levels: the difference between ``provided" and ``usable"

A TSA without service levels is like buying a generator that technically exists but may or may not turn on when the power goes out.

Service Level Agreements (SLAs) and key performance indicators (KPIs) are how the buyer turns "we'll help" into "you must respond within four hours to a P1 outage." The best TSA SLAs are not an encyclopedic wish list; they are a small set of measurable commitments tied to business-critical functions.

The seller will resist aggressive SLAs for the same reason any vendor resists them: penalties and escalation consume attention. Also, the seller's best people may have moved on to other internal projects. The buyer may feel like it is paying for a service provider with half a staff and a quarter of the urgency.

That friction is normal. The way through it is to anchor SLAs to what truly matters: payroll, billing, cash collection, order processing, cybersecurity, and core reporting. If everything is "critical," then nothing is.

Pricing: cost reimbursement, markups, and the awkwardness of being someone's customer

TSA pricing often starts with "seller's cost plus a markup." The markup might be meant to cover overhead, risk, or simply to motivate the seller to actually perform. Sometimes the TSA is priced at "cost" with minimal markup because the seller is trying to avoid being accused of extracting value post-close.

In practice, TSA pricing becomes contentious because "cost" is slippery. Does it include allocated corporate overhead? Does it include depreciation of shared systems? Does it include third-party vendor pass-through costs? What happens when usage is higher than assumed?

A good TSA treats pricing like a utility bill:

- Define the rate basis (FTE time, usage volume, fixed monthly).
- Define what's included (tools, licenses, subcontractors).
- Define how consumption is measured (tickets, users, transactions).
- Define a true-up mechanism if assumptions are wrong.

Yes, that is a true-up again—not the purchase price true-up, but the same idea: estimate now, reconcile later. TSAs have their own version of "your number vs. my number," and they will find you if you ignore them.

Governance: the TSA needs a steering wheel, not just an engine

The most common TSA failure mode is not that the seller refuses to provide services. It's that small issues accumulate into operational drag because no one owns the relationship.

TSA governance is the system for making decisions quickly:

- named points of contact on each side,
- a cadence of service review meetings,
- escalation paths for disputes,
- and a mechanism to approve changes (scope, pricing, term).

Without governance, the TSA becomes a daily negotiation conducted through angry emails and Jira tickets. With governance, it becomes what it should be: a temporary operating agreement with an end date.

The buyer should also track TSA performance like it tracks financial performance. Not because it's fun to score vendors, but because the TSA is often the most immediate operational dependency in the newly acquired business. If it falters, everything else is secondary.

Reverse TSAs: when the divested business becomes the vendor

In carve-outs and divestitures, there is a concept that surprises first-time deal teams: the reverse TSA.

A reverse TSA flips the direction. The divested business (now sold) continues providing services to the parent for a period of time. The parent might rely on the divested business for manufacturing output, specialized engineering support, certain IT tools, or customer service functions that weren't separated cleanly.

Reverse TSAs can be politically uncomfortable because they invert the power dynamic. The seller, who just "exited," may suddenly be a customer. The buyer, who just "acquired," may be stuck running legacy processes longer than planned because the seller needs continuity.

The operational lesson is the same: temporary arrangements have a way of becoming sticky. If the reverse TSA is not managed aggressively, the divested business can become a captive service bureau, distracting management from growth and making "integration" feel like running a museum.

The TSA trap: dependency that quietly hardens into architecture

There is a specific kind of integration failure that looks calm on the surface. The business runs. Invoices go out. Payroll hits accounts. Everyone congratulates the team for "stability." Meanwhile, the target is still living inside the seller's systems six months later, then twelve, then eighteen. The TSA gets extended again and again because the migration project keeps slipping.

At that point, dependency isn't temporary. It's architecture.

This is dangerous for three reasons:

1. *Cost creep.* TSA fees often rise after the initial period, or at least stop being "friendly." Even when pricing is stable, the buyer

pays twice: once for the TSA and once for its own build-out that never quite finishes.

2. *Risk concentration.* The target's operational resilience depends on a counterparty that no longer has skin in the game. If the seller reprioritizes, restructures, or suffers its own outage, the target inherits the blast radius.
3. *Strategic paralysis.* It's hard to modernize systems, change processes, or integrate data when the core platform isn't yours to change.

The fix is not willpower. It's planning with deadlines that bite: migration milestones, executive sponsorship, and a budget that treats exit as mandatory rather than aspirational.

The ``zero IT TSA" strategy: expensive before close, cheaper after

Some sellers push hard for minimal or no IT TSAs, especially in carve-outs. They don't want to run a shadow IT department for an asset they just sold. They don't want the security exposure of keeping a former division connected to their networks. They don't want stranded costs hanging around because the divested business still needs their tools.

A "zero IT TSA" strategy can be brilliant when it's feasible. It forces stand-alone readiness before closing: separate identity management, separate email domains, separate ERP access, separate cybersecurity monitoring, separate data hosting. It also forces the buyer to invest earlier and more aggressively.

That upfront investment can feel painful, but it buys something priceless: freedom. The buyer controls its own systems on Day 1, reduces ongoing dependence, and often lowers long-run integration risk.

The danger is pretending you can do "zero IT TSA" cheaply. If the plan is underfunded or rushed, you trade one risk (dependency) for another (operational failure). A missed payroll because the identity system wasn't ready is not a heroic story about "clean separation." It's a crisis.

People are part of the integration plan, even when the spreadsheet forgets

Integration is usually described in terms of systems and processes, but the business runs on human reflexes: the AP manager who knows which vendor always calls on the 28th, the IT lead who can restart a finicky server, the payroll specialist who remembers the one union site with special deductions.

Those people often sit inside the seller pre-close. After closing, they may be employees of the target, employees of the seller, or employees of a third party. Their incentives and loyalties can shift overnight. If they are covered under a TSA, they may be providing services to a business they no longer feel attached to. If they moved to the buyer, they may suddenly lack access to systems still controlled by the seller.

A smart Day 1 plan treats key people like key assets:

- Identify critical roles tied to Day 1 operations.
- Secure retention where needed.
- Ensure access rights and authority are aligned with responsibilities.

This is not sentimentality. It's operational risk management. A TSA can define services; it cannot manufacture institutional memory.

Data, security, and the awkward reality of shared systems

TSAs often require continued access to shared systems, and shared systems are where privacy and security risks concentrate.

If the target continues using the seller's ERP, HRIS, or email environment, questions become immediate and concrete:

- Who can see what data?
- How is access segmented?
- How are logs handled and retained?
- What happens if there is a breach—who investigates, who notifies, who pays?
- How are employees onboarded/offboarded without leaving ghost access?

These are not purely technical concerns. They can become legal obligations (data protection laws, contractual confidentiality, industry regulations) and commercial disasters (customer trust evaporates fast when you explain your systems were "in transition").

A TSA that covers IT should treat cybersecurity as a first-class service with defined responsibilities. "Reasonable efforts" is not an incident response plan.

Integration vs. separation: two words, same sleeplessness

Buy-side integration and carve-out separation have different end states, but they share a daily reality: the business needs to run while you change the tires at highway speed.

In a classic integration, the buyer is moving the target onto its own platforms: finance systems, HR policies, procurement processes, reporting cadence, internal controls. In a carve-out, the buyer is building platforms that may never have existed in stand-alone form. Either way, the high-risk period is the overlap—when the old world still exists, the new world isn't fully ready, and everyone is depending on temporary workarounds.

That overlap is where TSAs are meant to help. But TSAs are not a substitute for an integration plan. They are a timer. They measure how long you can borrow someone else's infrastructure before the bill becomes unacceptable or the relationship frays.

How to make TSAs boring (the compliment you want)

A TSA doing its job should fade into the background. No drama. No emergency extensions. No weekly escalations. Just steady service until the buyer no longer needs it.

That "boring TSA" outcome tends to come from a few habits:

- *Treat the TSA like a commercial contract, not a courtesy.* Define scope, SLAs, and escalation paths with the seriousness of a customer contract, because that's what it is.
- *Build an exit plan that is more than a date.* An exit date without milestones is a wish. Milestones without owners are a rumor.
- *Assume incentives are misaligned and design around it.* The seller will not prioritize your future. Your contract and governance must.

When the lights stay on after closing, nobody throws a party. They should. It means the deal team didn't just buy a company. They kept it alive while it changed skins.

6.3 Skin in the Game: Management Rollovers

A clean exit is a beautiful lie. In most founder-led deals—especially sponsor-backed ones—the buyer doesn't just want the assets and the customer list. The buyer wants the people who know where the bodies are buried, where the margins are hiding, and which "one-time" expense somehow shows up every quarter.

That's the real reason management rollovers exist. They're not a symbolic pat on the back. They're a practical tool for turning a seller who just took chips off the table into an owner who still cares what happens after the closing dinner.

Rollover equity, translated into plain English

A management rollover is when the founder and/or key managers reinvest part of their sale proceeds (or keep part of their existing equity) into the company's equity after the transaction. Instead of taking 100% cash and walking away, they keep an ownership stake in the post-close entity.

The pitch is usually delivered with a grin: "You get a second bite of the apple."

The mechanics are less poetic. The rollover changes three things at once:

- *Cash at closing.* Every dollar rolled is a dollar not wired to the seller that day.
- *Incentives post-close.* The manager now has something valuable that can grow—or get diluted, trapped, or wiped.
- *Power dynamics.* The founder goes from being the sole boss to being a partner with a boss, usually under a new governance regime designed for control and exit discipline.

And because it changes all three, it deserves the same level of attention as the headline price. A sloppy rollover can feel like a clever alignment mechanism right up until the moment the founder realizes they sold control and bought a minority security with terms they don't actually understand.

Why buyers push for it (and why they sometimes overdo it)

Sponsors love rollover equity for two reasons: economics and psychology.

Economically, rollover equity reduces the cash check the buyer has to write. In a leveraged buyout, where the sponsor is layering debt on top of equity, every dollar of rollover can lower the sponsor equity requirement. That can juice returns. It can also make a deal feasible that otherwise wouldn't fit the fund's underwriting.

Psychologically, rollover equity is a signal. If management is willing to bet on the future, the story goes, the future must be bright.

Signals are useful, but they are not magic. A founder can roll because they believe in the business, because they want to please the buyer, because they're told it's "market," or because their banker quietly hints that refusing will spook the sponsor. A rollover is not always a clean indicator of confidence. Sometimes it's just the toll to cross the bridge.

The buyer's temptation is to treat rollover as a universal good: more rollover equals more alignment. That can backfire. Overly large rollovers can leave founders feeling like they didn't actually get paid for the risk they already took, especially if the new equity comes with restrictions and preferences that make it behave less like "ownership" and more like a performance hostage.

Why sellers agree (even when they swear they're tired)

Founders don't roll because they're bored. They roll because the deal rewires their risk in a way that can be attractive.

Before a sale, a founder often has most of their net worth tied to one illiquid asset: the business. After a sale, they can take meaningful cash off the table and still keep upside exposure. It's concentration risk turned into a barbell: cash safety on one side, equity upside on the other.

There's also a pride factor that deal people underestimate. Many founders don't want to be remembered as someone who "sold out." They want to sell, keep building, and prove that the business wasn't just a lifestyle company that happened to catch a bidder at the right time. Rollover equity lets them stay in the arena—with a scorecard.

The catch is that the arena has different rules now.

The two rollovers: ``paper rollover" vs. ``new money"

Not all rollovers are created the same way, and the method matters for taxes, documentation, and leverage in negotiations.

Paper rollover (often called a "roll" or "rollover" in the narrow sense) is when the seller exchanges existing equity in the target for equity in the new holding company or buyer entity. Instead of cashing out that slice, they swap it for a new security.

New money rollover is when the seller receives cash for all their equity, then reinvests some of that cash into the buyer's equity (often at closing). Economically it can look similar, but legally and tax-wise it can be different. It also feels different: new money makes the reinvestment choice explicit, and sometimes the buyer likes it because it looks like a fresh vote of confidence.

In practice, deals can combine both. The important part is not the label; it's what the seller ends up owning after closing and what rights attach to it.

The cap table is where ``alignment" gets edited

The most seductive rollover conversation is the simplest one: "You'll keep 20%."

That sentence is almost meaningless without the rest of the cap table. Twenty percent of what class? With what preferences? With what dilution protections (or none)? With what distribution waterfall?

A manager can own 20% of the equity and still have very little economic participation if the sponsor structure has a large preferred return, aggressive management fees charged at the portfolio-company level, or multiple layers of instruments with senior claims. Conversely, a smaller percentage of a simpler structure can be worth far more.

The rollover negotiation is not just about percentage. It's about *where in the stack* that percentage sits.

Liquidation preferences: the clause that turns ownership into a waiting room

Many sponsor-backed structures use preferred equity for the sponsor and common equity for management. Preferred equity typically has a liquidation preference: a right to get paid back (sometimes with a return) before common equity receives proceeds.

A simple 1x non-participating preference means the sponsor gets its money back first, then the remaining proceeds are split. A participating preference can allow the sponsor to get paid back first and then still participate in the remaining proceeds. Add a preferred return hurdle and the common can sit underwater for a long time.

None of this is inherently unfair. It reflects risk allocation: the sponsor is putting in cash, and it wants priority. The issue is *expectation management*. Founders often hear "you'll still own meaningful equity" and imagine their ownership behaves like the stock they used to own—where every incremental dollar of value lifts their stake proportionally.

Preferred structures can break that intuition. They can turn the common into a call option that only starts paying after a threshold is crossed. Options can be wonderful when the threshold is realistic. They can be demoralizing when the threshold is a mountain and the management team is expected to climb it while being told they're "partners."

A rollover that is meant to align incentives should not accidentally convince management that their upside is an illusion. Motivation is not a line item you can accrue; it either shows up to work or it doesn't.

Governance: the founder keeps equity and loses the steering wheel

Rollover equity often comes with a psychological bait-and-switch. The founder remains an owner, but control typically shifts to the sponsor or strategic buyer through board rights, protective provisions, and vetoes over major decisions.

This is not a surprise to anyone who reads the documents. It is still emotionally jarring in practice.

Founders who were used to deciding on a whim now operate inside a governance system:

- budgets must be approved,
- acquisitions must fit a model,
- hiring the CFO might require a board process,
- and "we've always done it this way" becomes evidence for why it should be changed.

The rollover can soften that by making the founder feel invested, but it doesn't change the power structure. The founder is now an investor-manager hybrid. That can be energizing for the right personality. For others, it feels like becoming a well-paid employee in the company they built, which is a strange kind of grief.

The practical point: governance terms are part of the rollover economics. If the buyer has unilateral rights to force a sale (drag-along), change the capital structure, or issue new equity, then the value of the rollover depends heavily on the buyer's future decisions.

Minority protections: the difference between ``equity" and ``a story about equity"

Most management rollover holders are minority investors. Minority investors live and die by the fine print.

Some protections are standard and reasonable:

- information rights (financial statements, budgets),
- tag-along rights (ability to sell alongside the majority),
- preemptive rights (ability to participate in future issuances),

- and constraints on amendments that uniquely harm the minority class.

But "standard" is not the same as "automatic." Many deals offer management a meaningful economic stake but very limited control rights, on the theory that governance complexity slows down decision-making. Sponsors also want flexibility to refinance, recapitalize, and execute add-on acquisitions without running a shareholder town hall.

That tension can be managed if it's acknowledged. It becomes corrosive when management believes they are co-owners in the ordinary sense and then discovers they are along for the ride in the legal sense.

A rollover should be presented honestly: not as continued autonomy, but as continued exposure to upside.

Vesting, forfeiture, and the moment equity starts behaving like compensation

Another subtle shift happens when management equity is subject to vesting or forfeiture provisions. A founder might roll equity and also receive new equity incentives (options, profits interests, restricted units). Those incentives often vest over time and can be forfeited if the manager leaves or is terminated for cause.

Vesting is a retention tool. It can also feel insulting to someone who just sold the company they built. The emotional reaction is predictable: "You're making me earn the equity I already owned?"

The buyer's view is equally predictable: "We're paying you for the asset; the incentive is for the future."

Both views can be true. The key is not avoiding vesting; it's separating the concepts cleanly:

- Rollover equity is usually treated as *property* the founder already owns and is exchanging into the new structure.

- New incentive equity is usually treated as *compensation* for future performance and retention.

When those get blended—for example, when a portion of rollover is subject to forfeiture—the rollover stops feeling like investment alignment and starts feeling like a retention hook disguised as ownership. Sometimes that's commercially necessary. It should never be casually done.

Drag, tag, and the reality of the exit clock

A sponsor-backed deal has an exit plan baked into it, even when nobody says the quiet part out loud. That means management's rollover equity is typically illiquid until a future sale, recapitalization, or IPO. It may not be transferable. It may be subject to buyback rights.

Drag-along rights allow the majority to force the minority to sell in an exit. From the sponsor's perspective, drag is essential: it prevents a small holder from blocking a transaction. From management's perspective, drag can be comforting (you won't be stranded) or unsettling (you can be forced to sell on someone else's timetable).

Tag-along rights allow minority holders to participate if the majority sells. They are a classic fairness mechanism, but their precise drafting matters. A tag right that excludes certain types of transfers can be a tag right that rarely tags.

The important practical takeaway: rollover equity often comes with less control over *when* you get liquidity than founders expect. They have upside, but they are not holding the steering wheel of the exit.

Taxes (briefly, because they can ruin a good deal fast)

Tax treatment can vary widely based on structure, jurisdiction, and the form of equity received. The high-level point is that "rollover" does not automatically mean "no tax."

Some rollovers are structured to qualify for nonrecognition treatment, where the seller exchanges equity for equity and does not recognize gain immediately (subject to a long list of conditions and limitations). Other rollovers are effectively a taxable sale followed by a reinvestment, which can create tax at closing even though some value is being reinvested.

The details are the domain of tax advisors, but a founder should internalize one rule: never assume the rollover is tax-free simply because it is called a rollover. The label is marketing. The structure is what matters.

And taxes are not just about rate; they're about timing. A rollover that creates tax without cash to pay it is a bad surprise to discover after the wire hits.

The rollover as a behavioral contract

It's easy to describe rollover equity as "alignment." It's more accurate to call it a behavioral contract.

The buyer is saying: "Act like an owner after you've already been paid."

The manager is saying: "Treat me like a partner even though you control the board."

If those two sentences are true in practice, rollover equity can be powerful. Founders stay engaged, the sponsor gets a motivated operator, and the business benefits from continuity. The best outcomes often come from a founder who genuinely wants to keep building and a buyer who genuinely respects the founder's operating instincts.

If those sentences are false, rollover equity becomes a slow grind. Management feels like it took less cash for paper that doesn't matter. The sponsor feels like it gave equity to someone who no longer has hunger. The cap table becomes a list of resentments.

The surprising part is how often the economics are not the real issue. The real issue is expectations: what control looks like, what upside is realistic, what support the buyer will provide, and what happens if the relationship sours.

A practical way to sanity-check a rollover

Before agreeing to a rollover, management should be able to answer a few blunt questions without hand-waving:

- What exactly am I receiving (class of equity), and what sits above it?
- Under a mediocre exit, a good exit, and a great exit, what do I actually take home?
- Who can dilute me, and under what circumstances?
- What rights do I have to information, and how often will I get it?
- What happens if I leave—voluntarily, involuntarily, or due to disability or retirement?

These are not "lawyer questions." They are wealth questions. If the rollover is meaningful, the answers should be clear enough that a founder can explain them to a spouse, a financial advisor, or their future self at 3:00 a.m. when they're wondering why they agreed to a preference stack that turns their "equity" into a mirage until the sponsor doubles its money.

A good rollover doesn't just keep skin in the game. It makes sure the skin still has nerve endings.

6.4 The High-Wire Acts: Carve-Outs and Distressed Sales

Some deals feel like buying a house: you tour it, you inspect it, you negotiate what stays, you close, and then you argue about the scratch on the hardwood. Carve-outs and distressed sales are not that. They're closer to relocating a living organ—fast, delicate, and with a very real chance someone wakes up mid-procedure.

The common thread is that "the business" is not sitting there as a neat, self-contained box. In a carve-out, the thing you're buying has been living inside a larger body for years, sharing blood supply and nervous system. In a distressed sale, the thing you're buying may be technically alive but metabolically failing, with creditors, courts, and deadlines acting like gravity. In both cases, the usual M&A comfort blankets—time, clean financials, cooperative counterparties—get yanked away.

Carve-outs: buying something that was never meant to stand alone

A carve-out is the sale of a division, product line, geography, or subsidiary that has been operating as part of a larger company. The asset may have customers, employees, and revenue, but it often doesn't have its own full set of systems, contracts, or corporate functions.

If you ask, "Does it have a CFO?" the answer might be "sort of." If you ask, "Does it have its own ERP?" the answer might be "it uses the parent's." If you ask, "Who owns the customer contracts?" the answer might be "the parent does, but everyone knows those customers are really yours." "Everyone knows" is not a contractual term, and it does not survive closing.

The buyer's job in a carve-out is to turn an embedded business into a stand-alone business without breaking the revenue engine. The seller's job is to remove the asset without leaving behind a tangled mess of obligations, liabilities, and stranded costs. Both jobs are harder than they look in the deck.

The financial statements problem: when ``carve-out EBITDA" is a narrative, not a number

Carve-out financials are often built from allocations. Shared corporate costs get spread across business units. Centralized procurement discounts get attributed based on usage. IT costs are allocated using headcount or revenue or whatever proxy the finance team can defend with a straight face.

Allocations are not inherently wrong. They're just inherently arguable.

That becomes critical when a buyer tries to answer basic diligence questions:

- What does the business really cost to run on its own?
- What costs will disappear when separated, and what costs will reappear as stand-alone overhead?
- How much margin is "real," and how much margin is subsidized by the parent's scale?

In a carve-out, the Quality of Earnings work is often less about catching fraud and more about translating an internal accounting construct into something that can survive outside the parent. The buyer isn't just underwriting performance; it's underwriting *separability*.

Stranded costs: the ghost that haunts both sides

Stranded costs are the expenses that remain with the seller after the carve-out because the parent can't downsize shared functions quickly or efficiently. Think: the data center lease, the HR systems contract, the tax department headcount, the procurement platform subscription. When the business unit leaves, the parent loses some demand, but the fixed cost stays.

Sellers hate stranded costs because they make the remaining company look worse after the divestiture. Buyers should also care, because stranded cost pressure shapes seller behavior. A seller under pressure to eliminate stranded costs may:

- push aggressively for a short TSA term,
- refuse to provide certain services,
- demand high TSA pricing to compensate for temporary overhead,
- or insist the buyer take on contracts and people the buyer doesn't actually want.

A carve-out negotiation often has an unspoken secondary bargaining table: not just price for the asset, but who eats the ugly transitional economics.

The TSA becomes the deal, not a schedule in the back

In a carve-out, the Transition Services Agreement is not a convenience. It is frequently the operating plan for the first 6–18 months. The buyer is "owning" the business while renting the seller's back office, systems, and sometimes even physical infrastructure.

That sounds manageable until you remember what services are typically shared in a corporate environment:

- identity and access management (logins, permissions, security groups),
- cybersecurity monitoring and incident response,
- payroll and benefits administration,
- finance close and reporting systems,
- procurement tools and vendor onboarding,

- customer support platforms and telephony,
- and data warehouses that feed everything from pricing decisions to regulatory reporting.

If those services wobble, the carve-out doesn't just experience "integration friction." It experiences operational failure: missed invoices, missed payroll, shipment delays, and unhappy customers who do not care that you are in the middle of a separation plan.

The counterintuitive truth: the buyer can win the price negotiation and still lose the deal if the TSA strategy is wrong. In carve-outs, operational continuity is an economic term.

Too much TSA vs. too little TSA: the carve-out version of oversteering

There is a balancing act that never looks balanced while you're doing it.

If you rely too heavily on a TSA, you risk building a business that can't stand on its own. Deadlines slip because "the seller is still doing it." Data migration gets postponed because "the old system still works." The TSA quietly extends, the costs creep, and the carved-out business remains psychologically and operationally dependent.

If you rely too little on a TSA, you can end up trying to build a standalone company in a timeframe that ignores physics. ERP implementations, payroll transitions, cybersecurity tooling, and new vendor contracts are not weekend projects. When they're rushed, they break in ways that can take months to repair.

Carve-out success usually comes from being decisive about what must be independent on Day 1 versus what can be transitional. The trick is not being heroic; it's being realistic. Make the "must be ready" list short and brutal, and then actually fund it.

The ``zero IT TSA" fantasy (and when it's actually a strategy)

Some sellers push for "no IT TSA" because IT is where risk and cost live. Keeping a sold business connected to your network is a cybersecurity nightmare. Running IT services for a business you no longer own turns your IT team into an outsourced vendor overnight, usually without extra staff or gratitude.

A "zero IT TSA" approach can work when the buyer is willing to spend early and the target's systems can be separated cleanly. It is most plausible when:

- the carve-out already has semi-independent systems,
- the buyer has a strong platform to migrate onto,
- data can be extracted without contractual or regulatory landmines,
- and leadership treats the separation as a core workstream, not a side quest.

It fails when it's used as a negotiating slogan rather than an engineering plan. "No IT TSA" without a credible cutover and security model is how you end up with shadow systems, spreadsheets replacing controls, and an outage that makes everyone nostalgic for the TSA they refused to sign.

Contracts and consents: the quiet cliff edge

Even if the operational pieces are solvable, carve-outs often run into an unromantic legal constraint: many contracts cannot be transferred without consent.

Customer agreements, supplier contracts, software licenses, leases, permits, and financing arrangements may be held by the parent entity. The carved-out business may be "the one using them," but the counterparty's signature may be required to assign them.

This becomes a race with two clocks:

- the deal clock (closing date),
- and the consent clock (how long counterparties take to review and sign).

When consents don't arrive, parties rely on transitional arrangements: subcontracting, pass-through services, or interim licensing. Those workarounds can keep the business running, but they also create compliance and relationship risk. Customers do not enjoy learning, after the fact, that their contract was being "performed through an arrangement" while paperwork caught up.

A carve-out buyer should treat contract mapping and consent planning as core diligence. If the revenue depends on contracts that can't transfer, you are not buying a business. You are buying a persuasion campaign.

Now add distress: buying a melting ice cube with a stopwatch

Distressed M&A has its own gravity. The seller may be insolvent or on the edge. Liquidity is tight. Vendors are nervous. Employees are reading rumors. Customers are making contingency plans. Every week that passes can shrink the business you think you are buying.

That's why distressed deals reward speed. But speed changes the rules of diligence. You don't get to know everything. You get to choose what you must know, what you can infer, and what you will price as risk.

Distressed deals also change the counterparty set. In a normal sale, you negotiate primarily with the seller. In distress, you negotiate with:

- secured lenders who may effectively control the process,
- unsecured creditors who may object,
- committees who hire their own advisors,
- and sometimes a court that must approve the transaction.

The business is still there, operating. But the sale process is a legal and financial machine with its own rhythm, and it does not wait for your integration plan to feel comfortable.

Section 363 sales: why buyers show up even when it's chaotic

In the U.S., a common mechanism for distressed asset sales is a sale under Section 363 of the Bankruptcy Code. Buyers like 363 sales for a simple reason: they can offer a path to purchase assets "free and clear" of certain liens and claims, subject to court approval and the specifics of the sale order.

That phrase—"free and clear"—is catnip. It suggests a reset button.

The reality is more nuanced. The buyer can often avoid taking on many legacy liabilities, but not all risks evaporate. There can be successor liability theories, regulatory obligations, employee-related liabilities, and practical realities like customer disputes that follow the business whether or not they follow the legal entity.

Still, 363 can provide something that normal M&A struggles to provide when a company is collapsing: *transaction certainty*. Court-approved processes, bid procedures, and protections for good-faith purchasers can make it harder to unwind the sale after the fact, especially when parties fail to obtain a stay pending appeal.

If normal deals feel like negotiating in private, 363 deals feel like negotiating on stage. You are not only convincing a seller. You are convincing a system.

The bid procedures game: stalking horses, topping bids, and strategic discomfort

Distressed sales often involve bid procedures designed to maximize value for the estate. A "stalking horse" bidder sets the initial bid and terms, then other bidders can attempt to top it. The stalking horse may receive protections like breakup fees or expense reimbursement for setting the floor.

For buyers, this is both opportunity and irritation:

- Opportunity because you can acquire assets quickly, sometimes at an attractive valuation, with structured protections.
- Irritation because you may spend serious time and money to be the stalking horse, only to be outbid at the auction after you did the heavy lifting.

The stalking horse decision is a strategic one: is your bid primarily about price, or about terms, speed, and certainty? In distressed deals, terms can be value. A higher nominal bid that is conditional and slow may lose to a slightly lower bid that closes cleanly.

Diligence under time pressure: you won't get everything, so choose wisely

Distressed diligence is triage. The buyer should focus on what can kill the deal post-close:

- liens and title to key assets (can you actually own what you're buying?),
- critical contracts and whether they can be assumed/assigned,
- customer concentration and churn risk under distress,
- environmental and regulatory liabilities that can attach regardless of structure,
- and the operational capability to run the assets immediately after close.

Notice what's not on that list: a perfect understanding of every working capital line item. In distress, the business may be burning cash so quickly that the elegant mechanics of a traditional true-up are less relevant than the question of whether the assets can generate cash next month.

That doesn't mean price adjustments disappear. It means the buyer's risk management shifts from "make the purchase price precise" to "make the acquisition survivable."

The sale order and APA: where lawyers and operators should actually be friends

In a 363 sale, the Asset Purchase Agreement (APA) and the court's sale order work together to define:

- what assets are included,
- what liabilities are assumed,
- what contracts are assumed and assigned,
- and what protections the buyer receives.

Operators sometimes treat these as legal documents that happen somewhere else. That's a mistake. The "assumed liabilities" schedule is an operations document. It determines what obligations the business will have on Day 1: employee arrangements, customer credits, warranty claims, service commitments, and vendor disputes.

A buyer that only underwrites the P&L and ignores assumed liabilities is buying a surprise generator. It will run at full capacity immediately after closing.

Integration in distress: you don't get a TSA, you get a handoff

TSAs exist when the seller is healthy enough—and motivated enough—to provide ongoing services. In distress, that is often not the case. The seller may not have the staff, the systems may be failing, and the focus may be on conserving cash and complying with process requirements.

So the buyer has to plan for a more abrupt transition:

- stand up payroll and HR administration quickly,

- ensure systems access and data capture immediately,
- stabilize key vendor relationships with new credit arrangements,
- and communicate with customers to prevent a revenue cliff.

The buyer in a distressed deal is often acquiring not just assets, but a rescue mission with a purchase agreement attached. The first weeks post-close can be the entire investment thesis: stabilize, retain, rebuild.

Carve-out plus distress: when complexity stacks, not adds

The scariest deals are not pure carve-outs or pure distress. They're both.

Imagine a parent company under liquidity pressure selling a division quickly. The division is deeply embedded in the parent's systems. The parent wants a short TSA, minimal ongoing obligations, and maximum cash proceeds. The buyer wants continuity, separability, and enough time to stand up independent infrastructure.

Now add creditor pressure and a ticking clock.

In that scenario, you should expect:

- aggressive timelines,
- incomplete data,
- limited seller support post-close,
- and a separation plan that must work even if cooperation deteriorates.

These deals can still be excellent. They are often where the best risk-adjusted returns live, precisely because they scare off less prepared buyers. But the edge comes from operational competence as much as financial engineering.

A buyer's survival mindset for special situations

Carve-outs and distressed sales reward a particular kind of discipline: accepting that you cannot eliminate uncertainty, only *choose which uncertainty you can live with.*

In a standard deal, you can obsess over precision: the perfect working capital definition, the cleanest covenant package, the most exhaustive disclosure schedule. In special situations, precision is still valuable, but robustness is priceless. Robustness looks like:

- a separation plan that assumes delays and still functions,
- an operating model that can run without the seller's infrastructure,
- a legal structure that is clear about what you are buying and not buying,
- and a leadership team that can make decisions quickly without waiting for perfect information.

These deals are called high-wire acts for a reason. The trick is not pretending the wire isn't there. The trick is bringing the balance pole: the right TSA strategy, the right diligence triage, and the right operational readiness to land the business on its feet when the paperwork stops protecting you.